T0332325

WHY NORTH
IS UP

WHY NORTH IS UP

MAP CONVENTIONS AND WHERE THEY CAME FROM

Mick Ashworth

Bodleian Library
UNIVERSITY OF OXFORD

First published in 2019 by the Bodleian Library
Broad Street, Oxford OX1 3BG
www.bodleianshop.co.uk

2nd impression 2022

ISBN 978 1 85124 519 2

Text © Mick Ashworth, 2019

All images, unless specified on p. 219, © Bodleian Library, University of Oxford, 2019

Mick Ashworth has asserted his right to be identified as the author of this Work.

Cover design by Dot Little at the Bodleian Library
Designed and typeset in 12 on 16 Fournier by illuminati, Grosmont
Printed and bound in Wales by Gomer Press Limited on 150 gsm Arctic matt art paper

British Library Catalogue in Publishing Data
A CIP record of this publication is available from the British Library

CONTENTS

ACKNOWLEDGEMENTS

I want to thank Jennifer for her enthusiasm for this project, her constant encouragement and support, and for reading and refining the text. Thanks also to our boys for their excitement about it and for keeping my feet on the ground.

I have greatly benefited from help and advice from many individuals but particularly want to thank David Fairbairn for his review of the text, and those who provided nuggets of specialist information: Stuart Caie, Steve Chilton, Catherine Delano-Smith, Richard Oliver, Ed Parsons and Adrian Webb. Also many thanks to Nick Millea and the team in the Bodleian Map Room for their invaluable help in identifying maps from their collection to illustrate the book.

The book is dedicated to the memory of my friends Brian Miller and Paul Younger, who both loved books and maps and were always so willing to share their knowledge, wisdom, encouragement and friendship.

LATITUDE   **NORTH**   EQUATOR

TORRID ZONE

SOUTH POLE

MERIDIAN

EQUINOX

WEST

EAST

NORTH POLE

ZENITH

NADIR

LONGITUDE

Scale of Miles.

OCEAN-CHART.

INTRODUCTION

In Lewis Carroll's poem *The Hunting of the Snark*, published in 1876, a ship's crew were delighted to be provided with a map for their potentially hazardous voyage. Their main delight, though, was in the fact that it was 'A map they could all understand... A perfect and absolute blank!' They had no time for map symbols – 'conventional signs' – and the complexities of lines of latitude and longitude – presumably because they just didn't understand them, or felt they didn't need to.

Maps hold a great fascination for many people – perhaps from an interest in a particular place, journeys they've made or wish to make or just for their appeal as graphic images. They feature a captivating blend of art and science, and their complexity and level of detail lead to a common assumption that they are always true and accurate. There may also be assumptions on the part of cartographers that maps are widely and easily understood. But is this the case?

Throughout the history of maps, methods have developed to allow features on the Earth to be mapped, and as these became established and more universally used they became accepted as conventions. These conventions – part of the 'language' of maps still widely used by map-makers – aim to make maps more consistent, understandable and useful for their users, and can greatly influence the way maps look and how they are interpreted.

OPPOSITE The blank Ocean Chart from *The Hunting of the Snark*, a poem by Lewis Carroll, 1876.

And yet they are perhaps not greatly understood by general map users. By exploring their origins and development – including the underlying structure of maps, their symbols, how the shape of the landscape is shown, how names and boundaries are treated and methods behind thematic and specialized maps – this book offers a greater understanding and appreciation of the cartographer's task and of how maps work.

Maps serve many purposes, and different conventions have developed to meet these needs, driven by technological developments – in map-making and other fields – by the needs of map users and by changes in the information and data available to cartographers. The need for maps and the detail which they can, and are expected to, contain has changed hugely. Religious beliefs and geographical speculation were important factors in medieval maps, for example, while accuracy and details of faraway places became more critical for exploration through the Age of Discovery and as maps became more widely used for military and administrative purposes. Today, users assume they can search online for just about any place in the world and see a map of it instantly.

General topographic maps, statistical and thematic maps, navigational charts, military maps and mobile mapping applications all need to be structured in a certain way in order to fulfil their purposes. They need to be drawn to scale, to represent the spherical Earth in two dimensions, to use a combination of identifiable point, line and area symbols and to show places and boundaries accurately and clearly. But in doing this, and because maps are often personal abstractions of the world, cartographers need to make difficult choices, and compromises are often made, with places being omitted, areas distorted, features emphasized, statistics (mis)interpreted and place names spelt in certain ways in order to present a particular view. For these reasons, caution sometimes needs to be exercised in interpreting maps.

Through our computers and smartphones, maps today play a greater role in our everyday lives than ever before, but are the mapping conventions that have developed over centuries still relevant? Some are so well entrenched that map-makers depart from them at their peril, but as they continue to respond to recent changes – particularly to digital technologies – new conventions are emerging. The mystique of maps, the control and consistency that have been exercised by mapping conventions, and perhaps even the general lack of understanding of maps displayed by the Snark hunters, appear to be declining as we interact with (or even create for ourselves) maps in new ways. New conventions may relate less to what maps show and how they look, and more to how we all use, contribute to and disseminate them day to day.

Cap d'Ortegal.

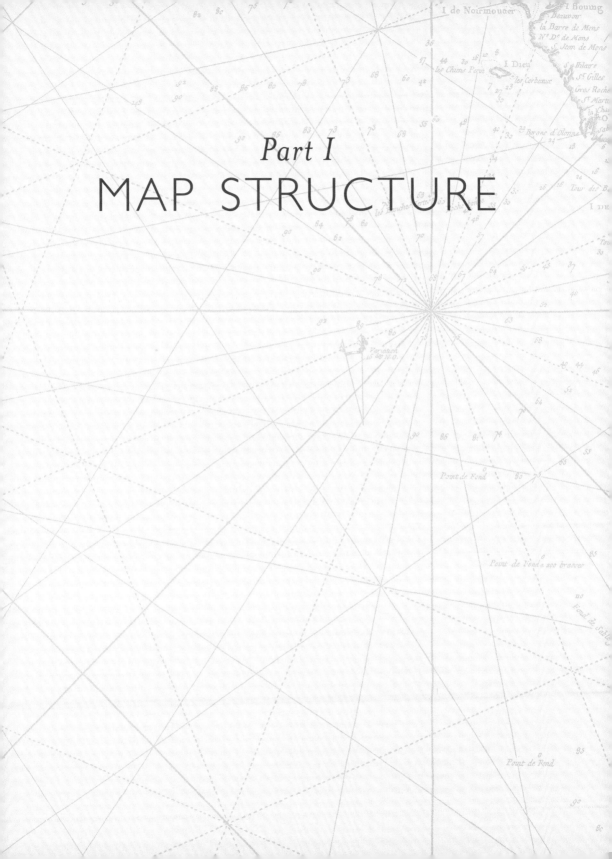

Part I

MAP STRUCTURE

EVRO

NORTH
This way up

In 1997 Ashley Sims produced a road atlas of Britain which promised to solve a long-standing problem with maps. His *Upside Down Map* of Great Britain claimed to allow 'easier and safer travelling from north to south' (fig. 2). Half of the atlas's maps are aligned traditionally with north at the top; the other half, including all their text, are turned around so that south is at the top of the page. Frantically turning maps around during a journey, to read place names easily and to be sure of reading right or left turns correctly, was to be a thing of the past.

This may seem a highly unconventional approach, but there is no rule that dictates which way up a map should be. The Earth doesn't have a top and a bottom. It is only since the nineteenth century that the convention of placing north at the top of maps has become universally accepted. Prior to this, maps were aligned in different ways, often following cultural and religious conventions rather than any geographical or scientific rules.

Throughout the history of cartography maps have been created in different orientations. The word 'orient' (as in to orient oneself or to align a map in relation to the real world) originates from the Latin *oriens*, meaning east – a nod to the fact that historical maps commonly had east at the top. What appears to be a common factor, however, is that the top of a map is

1 Compass roses became important practical devices for indicating orientation as maps were increasingly used for navigation and exploration. They also served decorative purposes – as on this chart of the Western Mediterranean by Bartolomeo Oliva from 1559.

perceived to be more important than the bottom. So maps were oriented towards those things deemed important to the prevailing culture and beliefs of the time – the residence of an emperor (who must be 'looked up to'), Mecca, the sunrise, the Pole Star, Paradise. Such perceptions are strong, and may be dangerous. It is sometimes thought that the adoption of the north-up convention has perpetuated a feeling that the northern hemisphere is in some way more important than the southern.

Ancient Mesopotamian maps were the first to take account of alignment. A stone-carved map of Nuzi (now Yorghan Tepe in Iraq) from *c.* 2300 BCE has east at the top and is the earliest map to give an indication of orientation. Ancient Greek and Roman maps generally favoured north orientation, although early maps of Rome carved on marble slabs – the *Forma*

2 Extracts from Ashley Sims's *Upside Down Map*, originally published in 1997. Through their different orientations, these pages promise to make the north–south journey to and from Durness much easier, but make no such provision for travelling west to east.

Urbis Romae, *c.* 203–208 CE – were oriented towards the south-east, in the direction of the city's most significant shrine. Ancient Chinese maps were commonly north-aligned, possibly reflecting the fact that the Chinese invented the magnetic compass during the Han dynasty.

The oldest surviving map with a realistic representation of Britain – an Anglo-Saxon world map, or *mappa mundi*, produced in 1025–1050 and known as the Cotton Map – has east at the top. Through medieval times this became an established approach in Europe with religious beliefs beginning to play a part. Christian maps, including perhaps the most famous map of the time, the Hereford Mappa Mundi, placed Eden, or Paradise – believed to be east of Europe – at the top, with Jerusalem in the centre. In contrast, Arabic maps, notably those of al-Idrīsī, placed south at the top – this being both the sacred direction of the Zoroastrian religion and the direction of Mecca from Baghdad and other significant cultural centres of the Middle East (fig. 3).

3 A sixteenth-century copy of the Arabic geographer al-Idrīsī's world map, from his *Entertainment for He Who Longs to Travel the World*, c. 1154. The shapes of Europe, Asia and northern Africa are remarkably familiar, although having south at the top, following Islamic tradition, is initially disorienting.

4 *McArthur's Universal Corrective Map of the World*, 1979. By breaking the north-up convention, and by centring the map on Australia rather than on Europe, the map-makers promised to change commonly held perceptions of Australia's place in the world.

The adoption of north as the predominant orientation was influenced by three key factors: the introduction of the magnetic compass to Europe from China in the twelfth century; the rediscovery in the fifteenth century of the map-making instructions of Claudius Ptolemy (*c.* 100–*c.* 170 CE) and Gerard Mercator's hugely important world map of 1569 (fig. 10). Certainly, as navigational techniques developed with the voyages of discovery, and as the more 'scientific' approach laid down by Ptolemy began to replace other mapping influences, it made sense to give north priority. Mercator's map, with north to the top and in a projection which revolutionized navigational methods, made it more difficult to go against this growing convention. The sharing of geographical knowledge between nations and across cultural divides, and the increasingly common use of sea charts produced by

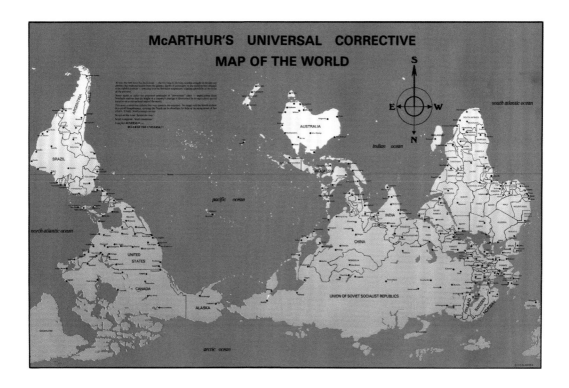

different nations following different conventions, highlighted a need for standardization.

However, rules are there to be broken and departures from the north-up convention are common. One notable subversion of the norm is *McArthur's Universal Corrective Map of the World* with south at the top (fig. 4). Published in Australia in 1979 it describes itself as part of a crusade to 'elevate our glorious but neglected nation from the gloomy depths of anonymity … to its rightful position – towering over its northern neighbours', and with Australia uppermost proclaimed an end to 'down under' jokes.

Whether for practical, political, religious or subversive purposes, we now have the capability with our smartphones and satnavs to orient a map display to our direction of travel at the touch of a button. North-to-the-top may not be such an important tenet for map users today, with Google Maps having turned the world upside down in more ways than one. Ashley Sims, once perhaps not taken too seriously, now looks to have been prophetic in his recognition of the need to choose a map's orientation to suit our immediate needs.

LATITUDE & LONGITUDE

Location, location, location

Today it seems that we always know where we are. The Global Positioning System (GPS) and similar satellite constellations provide data that allow us to pinpoint our location anywhere on Earth. Through our satnav systems, smartphones and computers we can position ourselves precisely in terms of latitude and longitude – the values on which the Earth's spherical geographical coordinate system is based. But the values defining our location – 51° 45′ 14.7″ N, 1° 15′ 14.6″ W for the Bodleian Library in Oxford, for example – could have been very different.

Latitude is defined relative to a plane which passes through the centre of the Earth, perpendicular to the axis around which the Earth rotates. This plane meets the Earth's surface to form the line known as the equator, measured as 0° latitude. Other planes parallel to the equator meet the Earth's surface to create a series of lines of latitude, or *parallels*. The angle between the equator and these lines define their latitude north or south of the equator. The poles are at the extreme limits of latitude, at 90° N and 90° S.

Similar imaginary lines, known as *meridians*, run north to south, perpendicular to the parallels, meeting at the poles. The angular measure of these meridians, relative to a defined reference point, is known as longitude. Whereas latitude is a natural coordinate, being measured in relation to the earth's axis and equator, longitude is a man-made coordinate, based on the arbitrary choice of a 'prime' meridian from which it is measured.

The concept of defining locations by a system of meridians and parallels was originally established by Ancient Greek geographers, mathematicians and astronomers, notably Dicearchus of Messina (360–c. 290 BCE), Eratosthenes of Cyrene (c. 275–c. 194 BCE) and Hipparchus of Rhodes (c. 190–c. 120 BCE). However, it is Ptolemy of Alexandria (c. 100–c. 170 CE) who is most commonly credited with establishing this method. In his *Geographia*, published c. 150 CE, he used a coordinate system to define the locations of all the known places in the world at the time. Although his own maps didn't survive, his system was 'rediscovered' and translated in medieval times and became the basis for mapping throughout the Arabic world and Europe, and later in China and Japan. Other than maps based directly on Ptolemy's work, the earliest 'modern' world

5 Ptolemy established the principle of defining location by latitude and longitude, although his mapping methods lay undiscovered, and unused, until the Middle Ages. Map of the world published by Donnus Nicolaus Germanus, c. 1460, based on Ptolemy's *Geographia*, c. 150 CE.

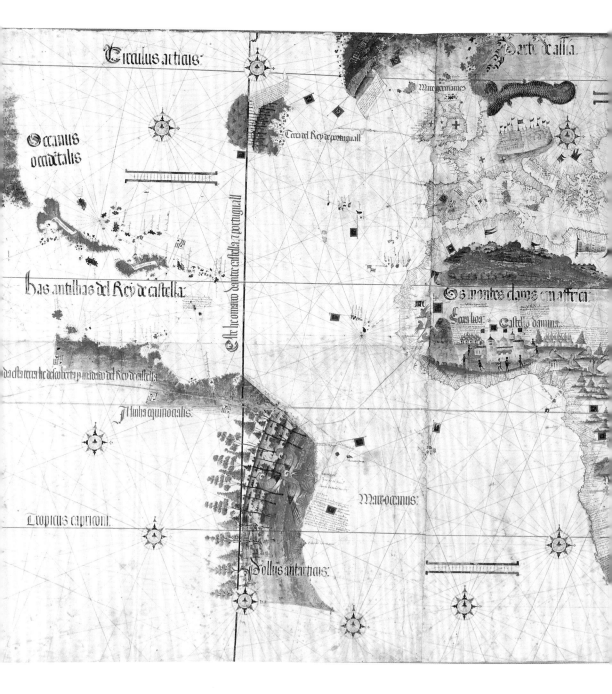

6 Named after the Italian who smuggled it out of Portugal in 1502, the *Cantino Planisphere* shows the meridian in the New World defined by the 1494 Treaty of Tordesillas – assigning lands to the west to Spain, and to the east to Portugal.

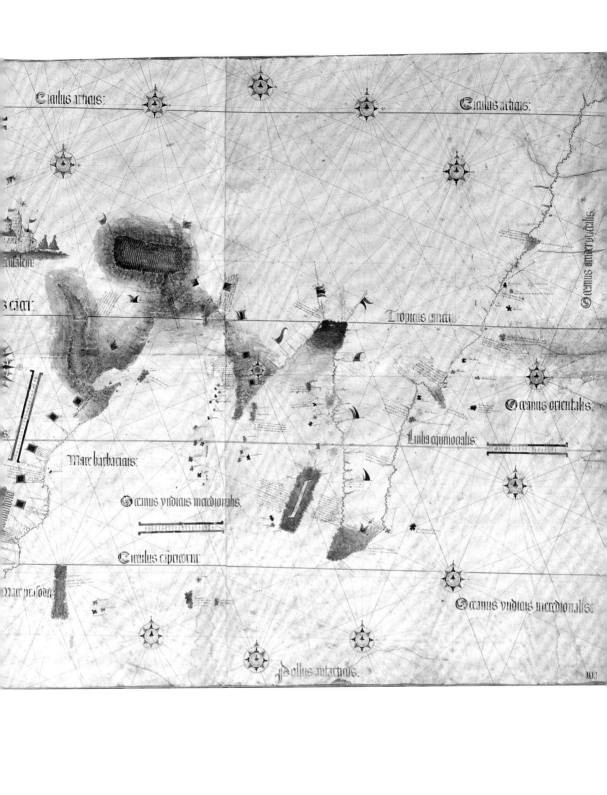

Circulus articus:

Circulus articus:

Oceanus amerovialis:

Tropicus cancri:

Oceanus orientalis:

Linha equinoctialis:

Mare barbaricus:

Oceanus yndicus meridionalis:

Circulus capricorni:

Oceanus yndicus meridionalis:

Mare prasodu

Pollus antarticus:

map marked with latitude and
longitude was by Heinrich Hammer
(Henricus Martellus Germanus), who
worked in Florence, in *c.* 1490.

With the point from which
longitude is measured being a
matter of personal choice – whether
to use a new position or to follow
any existing convention – Ptolemy
used the position of the Canary
Islands as his prime meridian. This
was because these islands were the
most westerly known point on the
Earth at the time. But as exploration
spread westwards during the Age of
Discovery, between the fifteenth and
eighteenth centuries, more appropriate and less arbitrary locations came
into use. Similarly, as more accurate
methods were developed for measuring latitude and particularly longitude
(whose precise measurement at
sea only became possible after the
creation of accurate chronometers
in the mid-eighteenth century –
notably John Harrison's Longitude

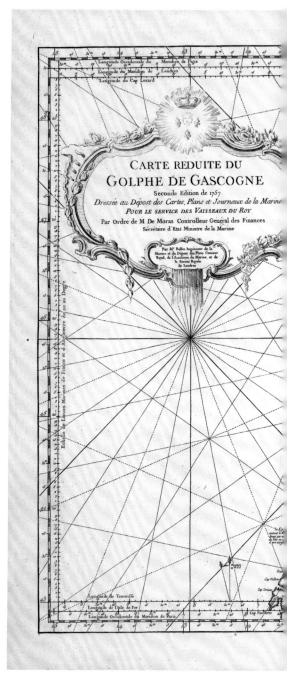

7 This map illustrates well, and tries to
account for, the use of different prime
meridians. It includes longitude measurements
from London, Paris, Lizard Point, Tenerife and
Isle de Fer. *Carte réduite du Golphe de Gascogne*,
Jacques Nicolas Bellin, 1757.

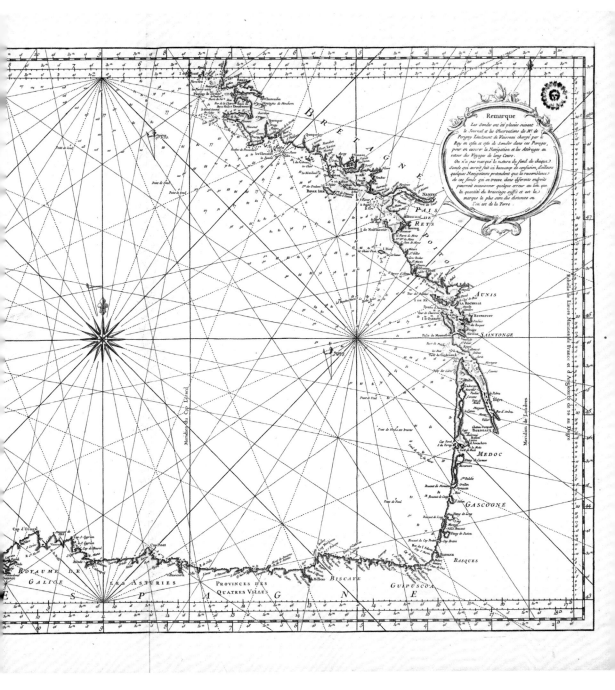

BRETAGNE

PAIS DE RETS

POITOU

AUNIS

LA ROCHELLE

ROCHFORT

I D'OLERON

SAINTONGE

BOURDEAUX

MEDOC

GASCOGNE

BAYONE

BASQUES

GUIPUSCOA

BISCAYE

PROVINCES DES
QUATRES VILLES

LES ASTURIES

ROYAUME DE
GALICE

Cap d'Ortegal

Cap Finis

E S P A G N E

Meridien du Cap Lizard

Meridien des Londres

NANTES

Remarque

Les Sondes ont été placées suivant
le Journal et les Observations de Mr. de
Perigny Lieutenant de Vaisseaux chargé par le
Roy en 1782. et 1783. de Sonder dans ces Parages,
pour en assurer la Navigation, et les Attérages au
retour des Voyages de long Cours.
On n'a pas marqué la nature du fond de chaque
Sonde qui aurait fait un beaucoup de confusion, d'ailleurs
quelques Navigateurs prétendent que la ressemblance
de ces fonds qui se trouve dans différens endroits
pourroit occasionner quelque erreur au lieu que
la quantité du brassiage suffit et est la
marque la plus sure des distances où
l'on est de la Terre.

A Paris chez les Libraires Marianade France et d'Anquetil de 1749 au Depôt

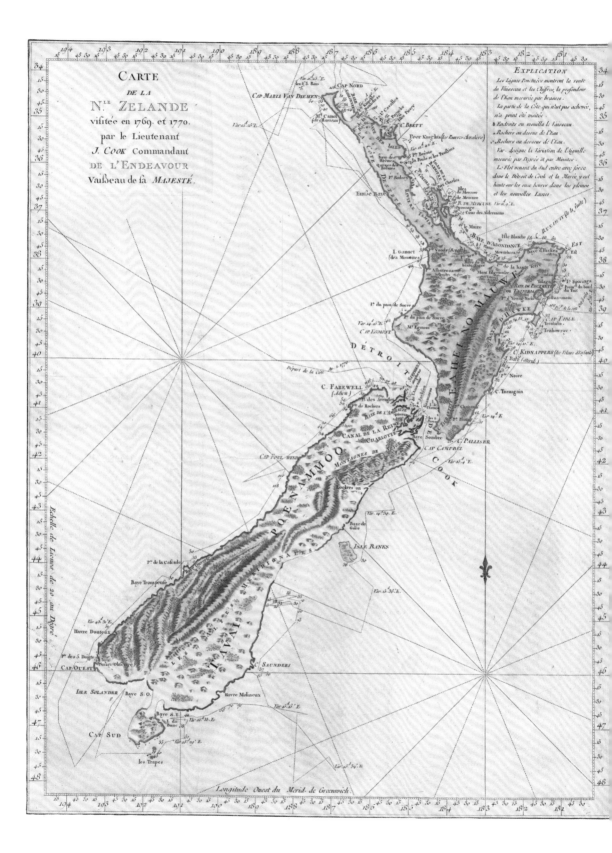

8 A French copy of Captain James Cook's map of New Zealand from the voyage of the *Endeavour*, 1769–70. The map is remarkably accurate in terms of latitude and the shape of the islands, and only about one minute in error in terms of longitude.

Prize-winning H4 chronometer of 1761 – the world map began to change beyond recognition, in terms of both content and positional accuracy.

Because many different nations were creating their own maps, numerous prime meridians came into use – more than twenty were

BREAKING THE CONVENTION

It can reasonably be assumed that geographical coordinates on maps will be accurate. But during the Cold War the USSR had a sophisticated programme of creating falsified maps of their own territory. Maps were altered to keep locations secret and to confuse potential enemies. This falsification included deliberately incorrect coordinates – maps showed real landscapes, but in totally wrong locations – possibly in the hope that enemy missiles would miss their real targets.

either in use or proposed by the late nineteenth century. Countries quite reasonably, often for political reasons, chose to centre their maps on a point within their own territory. Cadiz, Copenhagen, Jerusalem, Paris, Naples, Berlin, Ceylon, Rio de Janeiro, Tenerife, and even the home of a prominent eighteenth-century Japanese cartographer, all served as prime meridians for measuring location and compiling maps. This posed significant and obvious dangers to seafarers, who could be using charts produced in different countries whose coordinates disagreed.

It was only in 1884, at the International Meridian Conference in Washington DC, attended by representatives of twenty-five countries, that it was agreed that the meridian passing through Greenwich, London, should serve as the universal origin for measuring longitude. The aim of the Conference was to agree 'upon a meridian proper to be employed as a common zero of longitude and standard of time-reckoning throughout the world'. The result was the convention which we now take for granted, and which is built into many of our everyday activities whether we realize it or not – that of stating longitude in terms of degrees west or east of Greenwich.

MAP PROJECTIONS
Distorting the Earth

Peeling an orange is perhaps the most common illustration of a critical map-making problem. No matter how carefully removed, the peel cannot be laid flat as a continuous surface. There may be separate pieces, and there will be gaps. Translating – or 'projecting' – a near-spherical three-dimensional object into two dimensions is an issue that has faced cartographers since mapping began.

Map projections date back to Hipparchus (*c.* 190–*c.* 120 BCE), with more mathematical approaches to the problem being applied later, particularly by Marinus of Tyre (70–130 CE). But it was Gerard Mercator (1512–1594) who revolutionized the study and practice of projections with his hugely influential world map of 1569 (fig. 10). Since then, and particularly through the eighteenth century, numerous styles of map projection have been devised.

The term 'projection' became established as different methods of representing the Earth developed. The process can be visualized by imagining a light placed within a transparent globe to project the shape of the Earth's *graticule* (the lines of latitude and longitude – see p. 13) onto a piece of paper touching the globe. Depending on just where that light is positioned – at the centre of the globe, on the surface of the globe opposite the paper, or from a distance – the graticule will take on a certain pattern. That pattern will tell us something of the effects of that projection and how the globe's geography has been distorted. Meridians (lines of longitude) may not meet at the poles as they do in reality, parallels (lines of latitude)

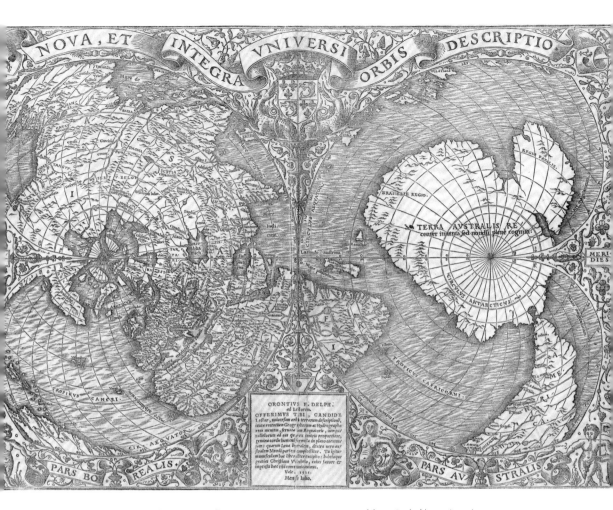

9 The world can be 'projected' in many ways to create a great variety of shapes. This double-cordiform map projection by Oronce Finé from 1531, controversial at the time, uses the equator to define the heart-shaped northern (left) and southern (right) hemispheres.

10 OVERLEAF Mercator's *Nova et aucta orbis terrae descriptio ad usum navigantium emendate accommodata* (*New and More Complete Representation of the Terrestrial Globe Properly Adapted for Use in Navigation*), 1569. Although the relative sizes of the continents are incorrect, this map revolutionized cartography and navigation.

may no longer be parallel to each other and the poles may appear as lines rather than points. The shape of the paper onto which the projection is made also determines the pattern. Projections are commonly described as *cylindrical* (when paper is wrapped around the globe), *conical* (a cone shape placed on the globe) or *azimuthal* (a flat piece of paper touching the globe at

a single point). One further complication is that map scale (see p. 35) is only correct at the point where the paper touches (or passes through) the globe. Projections are described mathematically as well as graphically, and there are infinite mathematically based projections. Whichever approach is taken, it is clear that to represent the Earth on a flat map something has to give. Shapes need to be distorted, areas adjusted, or angles and directions altered to make things fit.

Mercator's 1569 map illustrates this well. On his cylindrical projection, which is still in common use, any straight line drawn on the map represents a *loxodrome* or *rhumb line* – a line of constant compass bearing. This was revolutionary, and was of vital importance for navigation at a time when explorers were travelling further and further afield. However, this characteristic comes at a cost. The most common criticism of Mercator's projection is that it hugely distorts area. On modern versions, Greenland appears to be approximately the same size as South America when in reality the latter is almost eight times the size of the former. The true sizes of the continents were not as important then as being able to navigate easily, but today, seeing things on an 'equal-area' projection may be more important to allow true geographical comparisons to be made.

Such distortions, though, can also be used to good effect. Cartographers can now choose from many projections with different characteristics and often find a suitable compromise to keep distortions to a minimum. Projections are often classified as either *conformal* (retaining correct angles and therefore shapes), *equivalent* (representing areas correctly) or *equidistant* (with distances from a specific point being correct). The main purpose of a map will suggest to the map-maker which type should be used. A conformal projection, with correct shapes but incorrect areas, would not be appropriate for a map showing the density of a particular phenomena, for example.

All this is not generally in the public eye. However, in 1973 the German writer, film-maker and social activist Arno Peters brought such issues into

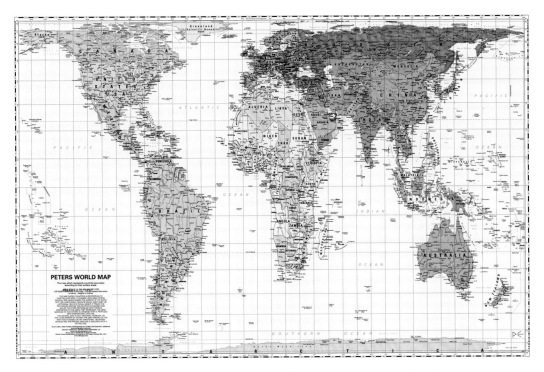

11 The *Peters World Map*, 1973, aimed to 'correct' the more common conventional map projections which distort the relative sizes of the world's land masses. Although it portrays countries and continents correctly in terms of area, their shapes are rather unconventional.

the spotlight by creating an equal-area projection which, he argued, righted the wrong of the way the developing world was commonly represented on world maps (fig. 11). Mercator's projection in particular was guilty (although more by accident than design) of exaggerating land areas belonging to the rich West, and Peters argued that its continuing use worked against the interests of the developing world. His map, although it was quickly identified by cartographers as a close copy of a projection developed by James Gall in 1855, was adopted by numerous international development agencies and the United Nations.

Although Peters's projection was not perfect – it is instantly recognizable by the strange elongated shapes of the continents, and it caused some controversy – it certainly made many people aware for the first time of the distortions inherent in map projections, and how they can influence our perceptions of the world.

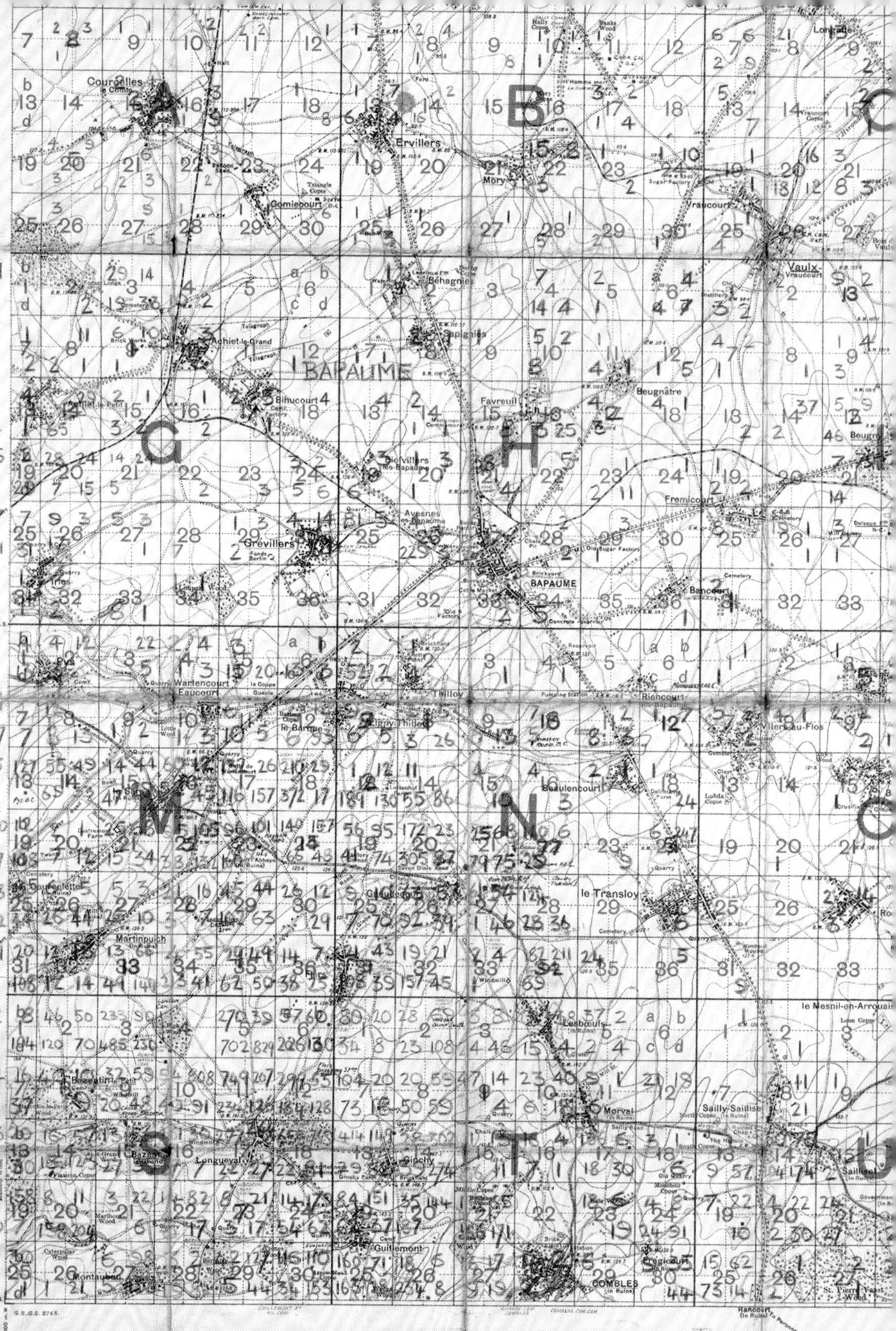

Courcelles-le-Comte
Ervillers
Mory
Gomiecourt
Vraucourt
Vaulx-Vraucourt
Béhagnies
Sapignies
Achiet-le-Grand
BAPAUME
Bihucourt
Favreuil
Beugnâtre
Beugny
Biefvillers-lès-Bapaume
Fremicourt
Avesnes-lès-Bapaume
Grévillers
Irles
Bancourt
Pys
Warlencourt
Eaucourt
Thilloy
Riencourt-lès-Bapaume
Villers-au-Flos
le Barque
Ligny-Thilloy
le Sars
Beaulencourt
Courcelette
Gueudecourt
le Transloy
Martinpuich
Flers
le Mesnil-en-Arrouaise
Lesboeufs
Bazentin-le-Petit
Morval
Sailly-Saillisel
Bazentin-le-Grand
Longueval
Ginchy
Guillemont
Montauban
COMBLES
Frégicourt
Rancourt
St. Pierre Vaast Wood
G.S.G.S. 2743.
PERONNE

GRIDS
Squaring up

It wasn't what the squares on the maps were intended for. During the First World War, the Red Cross kept records of deaths and looked after war graves. They used British topographic maps of France to record battlefield deaths, and, as was common on military maps, the maps included grids to help define and describe locations. Handwritten blue pencil marks were added to grimly record the body count for each 500-yard grid square on each map, the grid providing neat units by which to keep a tally. The maps provide a dramatic, and shocking, picture of the terrible loss of life (fig. 12).

However, grids were first introduced onto maps for more positive and constructive reasons and have become a conventional way of defining location – useful for the military to identify targets and set rendezvous points, but also for general users to share information about meeting points and define walking routes.

The Chinese used grids as long ago as the first century CE but the most remarkable Chinese gridded map to survive is the *Yu ji tu* (Map of the Tracks of Yu) created in *c.* 1137 (fig. 13). This map is engraved in stone and has a precise grid across its whole extent, with each grid square measuring 100 *li* or approximately 57 km (35 miles). The grid may have been the mathematical basis for the original construction of the map, but its prime

12 Maps such as this, based on a British military map of France, *GSGS 2743 1:40,000 Sheet 57c,* used the grid to help count, retrieve and rebury bodies of soldiers. This extract shows the area of the Battle of Delville Wood, 1916.

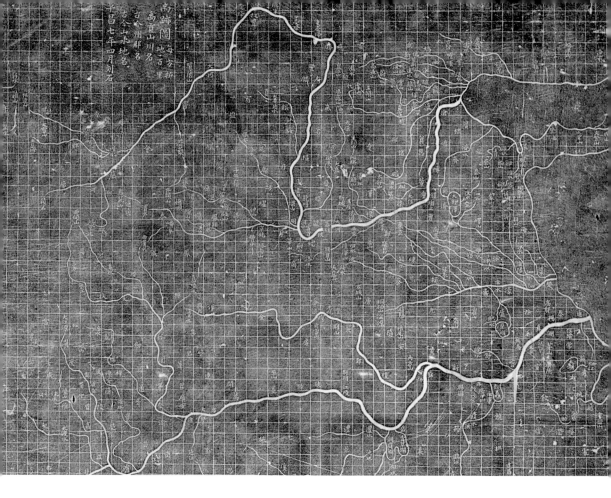

13 The *Yu ji tu*, *c.*1137, is one of the earliest maps to make use of an accurate, systematic grid. As well as indicating scale, the grid may have helped in the map's construction – including its remarkably accurate depiction of China's rivers.

purpose was to give an indication of scale. It may also have been to enable easy rescaling of the map – in a method similar to exercises in children's drawing books where a picture is copied square by square.

The Romans worked with grids in a different way. Their method of land division known as *centuriation* was based on a grid system. Later, in medieval Europe, Pietro Vesconte appears to have been influenced by the Chinese approach. His map of the Holy Land (*c.* 1320) carries a very similar grid to that of early Chinese maps, as do some navigational charts of the fifteenth century. The rediscovery of the work of Ptolemy in the Middle

Ages, and the advent of printed maps, led to an increasing use of grids. Ptolemy based his own reference system on lines of latitude and longitude (the graticule – see p. 22), a method still in common use today, particularly for small-scale atlas maps. But derivations of Ptolemy's work used different systems. Conrad Wolfhart, also known as Lycosthenes, used a world map based on Ptolemy, produced in Basle in 1552, to create what is the earliest surviving example of a grid-based map index. John Norden's 1593 map of Middlesex is among the earliest maps to use the now common method of an alphanumeric reference system to identify grid squares – with numbers along the top and bottom, and letters down each side (fig. 14).

Rectangular plane coordinate systems – simple grids based on a flat surface, using evenly spaced horizontal and vertical lines – became well established through the twentieth century and are the most common form of map grid. Users of Britain's Ordnance Survey (OS) maps will be familiar with the British National Grid. This was established in 1938 to ensure a uniform national reference system. Prior to this, maps of Britain used many different grids or no grid at all. They will perhaps be less familiar with the hundreds of similar local grid systems used throughout the world, and the Universal Transverse Mercator (UTM) system devised in the 1950s for use on global military maps. Such grids allow users to locate positions simply and accurately by quoting *eastings* – distances measured horizontally left to right – and *northings*, measured vertically bottom to top. These grids all need a point of origin from which distances are measured. The British National Grid has two. Its true origin is at 49° N, 2° W – off the south coast of England, south-east of Jersey. This is the approximate west–east centre of Great Britain and is the basis for the map projection on which OS maps are constructed. However, if grid coordinates were also based on this, some parts of England, the whole of Wales and most of Scotland, all west of this point, would have negative values. To eliminate the risk of confusion over similar positive/negative references (and perhaps to avoid unfortunate political connotations), a false origin was established west of the Scilly

Isles, 400 km west and 100 km north of the true origin. This means that all places within Britain have positive coordinates.

Grid references based on detailed grids of this type are of immense value, and grids have been an integral part of topographic maps for over a century. They have been particularly important and prominent on military maps (see p. 183), with positional accuracy seen as critical in carrying out successful campaigns. It seems a tragic irony that grids intended to ensure military success were ultimately used to count the cost of events during the First World War.

14 John Norden's *Myddlesex*, part of his *Speculum Britanniae*, 1593, helped establish conventions now widely followed, particularly on atlas maps – an alphanumeric grid, allowing places and features to be easily referenced, and an explanation of symbols, or legend (see p. 41).

PARTE OF HARTFORD

6

Potters barr

Southmyms

Durhams

Dancers hill

Kiekesend

ÿ folde — Hadley

High bernit

Elstre

Tatteridg

Northend

East Ende

Brownswel

Frythe

Drayers hill

Highwood hill

Milhill

Edgeworth

Dalis

Finchley

Hendon

Hendon house

Kinsbury

Preston

Ixenden

Brent street

Daleson hill

Wembly hill

Nesedon

Wilsedon

Kylbourn

W. Twyford

E. Twyford

Hangar Woode

Deane Woode

Paddington

Harleston grene

Westbourne

Hyde parke

E. Acton

W. Acton

Hanwell

Gunnersbury

Boston

Elinge

Turnhamgrene

Padingwik

Hospitale

Kensington

Brompton

Shelfey

Knights bridge

Hamersmyth

Parsons grene

Fulham

Marebone

S. Gyles

Westminster

LONDON

Southworke

Lambeth

Battersey

Clapham

Newington

Wunsworth

Barnes

Putney

Mortlake

Shent

SVRREY

Richmonde

Petersham

Worton

Coppermylle

 Buffleworth

Twiken hall pke

Twyckenham

Kenton bury

Hampton

New howse

Tuddington

Kingston

Hamton court

Ws. Mawlsey

ENFEILDE CHACE

Morthathe

Enfeyld house

Enfeylde

Ludgraves

Ponders ende

Grensfreete

Duranes

Theoball

Whitwebb

Waltham

Waltham crosse

FOREST OF WALTHAM

Waltham

Merchirch

SEPTEN. ORIEN.

OCCIDENS MERID.

Chigwell

Chmgford

PARTE OF ESSEX

Woodford

Walthamstow

Wansted

Layton

East Bernet

Whetstone

Fryarn Barnet

Friar Manner

Brumsfyld

Cony hatche

Muswell hill

Tottenham

Hollicke

Duccats

Tottenham highe. cr.

Harnesey

Cruch end

Ledgehill

Highgate

Chyldes hill

Hamsted

Belsarr

Chalcote

Halways

Kentish Towne

Islington

Canbury Pale

Newington grene

Newington

Shacklewell

Clapton

Hackeney

Mothers

Dom. Soy Sct.

S. Ric. Martin

Strete

Strete

Winchmorhill

Edm. streete

Edmondton

Wyerhall

Pymes

Burystreete

Kingsland

Hockesdon

Meres frete

Bush. hall

Stratforde

Stepney

Blackwall

Limehowse

Olde forde

Totten Courte

Clerkenwell

Shordich

Isle of doges ferm

Charleton

Derford

Grenewich

Part of Kente

Caracters diftinguishing the difference of places

☩ Market townes

◦ Parishes

⌖ Hamletes or villages

⊕ Howses & Palaces of Quen. Eli

△ Howses of Nobilitie

⊙ Howses of Knights, Gent. &c.

♜ Castles & fortes

⚜ Monafteries or religious howfes

♟ Bifhops Seas

† Hofpitales

✚ Places where battils have bene

✳ Decade places

♦ Lodges in forestes chafe &c.

✺ Mylles

Ioannes Norden Angl. defcripfit 1593

a b c d e f g h i k

SCALE
Size matters

In his short story *On Exactitude in Science*, published in 1946, Jorge Luís Borges describes an empire whose cartographers created 'a map of the empire whose size was that of the empire'. And in Lewis Carroll's *Sylvie and Bruno Concluded*, the character Mein Herr boasts of a map of his country 'on the scale of a mile to the mile'.[1]

Clearly a map at life-size remains in the realm of fiction. Maps are not true-to-scale pictures of the world but are representations at smaller scales. With reduced scale come limitations both on what it is possible to show and how it can be represented. The mapping process has three basic stages: defining the purpose of the map, deciding the area to be covered, and the choice of an appropriate scale. A map's scale – the ratio of the size of a feature on the map to the size of that feature on the ground – influences the level of detail that may be shown, the amount by which features need to be simplified (the process of generalization – see p. 91), and gives some indication of how accurate and comprehensive the map is.

These issues have been around since ancient times. A statue of Gudea, a ruler of the ancient Mesopotamian city of Lagash, from *c.* 2200 BCE has him holding a plan of a temple which includes a measuring rule representing its scale. The Ancient Greeks used precise units of measure

15 Christopher Saxton's *Map of Carnarvonshire and Anglesey*, 1578, part of his atlas of county maps of Britain, includes a scale bar of 10 miles. The decorative dividers seem to emphasize and promote the accuracy of Saxton's detailed survey of Britain.

for surveying purposes and passed on their methods to the Romans, whose *mensores* (measurers or surveyors) created detailed maps and plans. Accurate methods of measurement in China date back to the astronomer Chang Heng (79–139 CE) and Phei Hsui (224–271 CE). No maps from these periods survive, but remarkable maps from the Song dynasty illustrate the accuracy of ancient Chinese maps. The *Yu ji tu* map of *c.*1137 (see fig. 13) is a prime example, with a consistency of scale and accurate grid allowing easy measurement of distance.

The use of grids to indicate scale and help locate features continues today, but there are three principal means of representing scale on maps: graphical, mathematical and descriptive. Scale bars are graphic devices subdivided into relevant units of measure. Units can vary geographically and have changed over time – think of leagues, stadia, furlongs, chains, and even travel times. Medieval maps often carry several scale bars, giving the user a choice of units. Maps from this period also make great use of scale bars as decorative elements. Modern scale bars are much more basic, but serve the same purpose – allowing the user to relate distances on the map to actual distances on the ground.

The definition of the metre in 1791 led to the representation of map scale in mathematical terms. A *representative fraction* (RF) is the ratio of one map unit to the number of those same units on the ground. A scale RF of 1:50,000 means that a distance of 1 centimetre on the map represents a distance of 50,000 centimetres (500 metres) on the ground. On a smaller-scale map of 1:1,000,000, 1 centimetre represents 10 kilometres. The units of measure don't matter so long as they are consistent – one finger width on a 1:50,000 scale map represents 50,000 finger widths on the ground.

Map scale can sometimes be described in words. Perhaps the best-known example of this is the description of maps at 1:63,360 as being at a

16 The scale bar on this 1612 map of Virginia by its then governor John Smith is in leagues – equivalent to 3 miles, or an hour's walk. Full leagues are shown by ticks, and half leagues by the red and white subdivisions.

scale of 'one inch to a mile' (there being 63,360 inches in a mile). The old British Ordnance Survey maps of this scale are affectionately known as 'one-inch' maps. When these maps were superseded in the 1970s by a new series at 1 : 50,000, no neat statement was available – 'about one-and-a-quarter inches to the mile' doesn't trip off the tongue!

Whether maps are large- or small-scale can be confusing. There are no set definitions, but it is useful to be able to compare and recognize both types of map. A large-scale map should be thought of as a close-up view, showing large amounts of detail, and needing a large piece of paper to cover a defined area. A small-scale map of the same area is a more general, distant view, showing much less detail on a much smaller piece of paper.

Whichever group a map falls into, its scale will be determined by its purpose. Choice of an inappropriate scale may lead to the map becoming like Borges's 1 : 1 map, which was found to be 'useless', or Lewis Carroll's map, whose users changed their mind and declared 'we now use the country itself, as its own map, and I assure you it does nearly as well'.[2]

17 The first accurate survey of Mont Blanc takes a scientific rather than decorative approach to its representation of scale. It uses all three methods – representative fraction, scale bars and scale statement. Extract from *The Chain of Mont Blanc* by Anthony Adams-Reilly, 1865.

IE CHAIN OF MONT BLANC,

from an actual Survey in 1863_4,

by

A. ADAMS-REILLY, A.C., F.R.G.S.

SCALE 80.000 OF NATURE.

ENGLISH MILES.

0   1   2   3   4   5   6

ENGLISH FEET.

0 500 2000   5000   10.000   15.000   20.000   25.000   30.000   35.000   40.000

0   1000   2000   3000   4000   5000   6000   7000   8000 Mètres

0   1   2   3   4   5   6   7   8 Kilomètres

[.792 Inch to a Mile.]

Published under

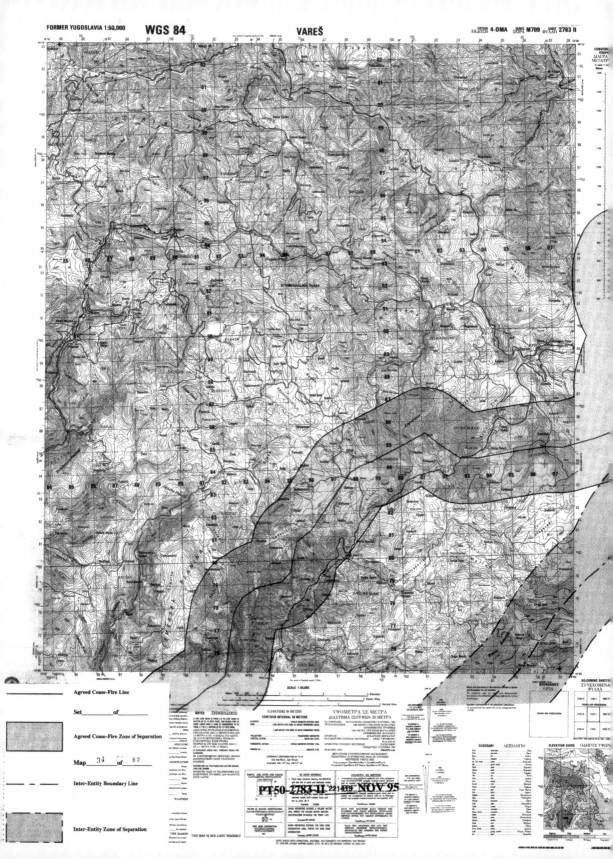

Agreed Cease-Fire Line

Set _____ of _____

Agreed Cease-Fire Zone of Separation

Map   34   of   53

Inter-Entity Boundary Line

Inter-Entity Zone of Separation

SCALE 1:50,000

ELEVATIONS IN METERS
CONTOUR INTERVAL 20 METERS

ΥΦΟΜΕΤΡΑ ΣΕ ΜΕΤΡΑ
ΔΙΑΣΤΗΜΑ ΙΣΟΫΨΩΝ 20 ΜΕΤΡΑ

PT50-2783-II   NOV 95

THIS MAP IS RED-LIGHT READABLE

LEGENDS
What does it all mean?

The ability of a map user to distinguish between, say, a symbol representing a church with a spire from that of a church with a tower may not seem too important. But in some situations – in particular during military operations or when navigating at sea – the misinterpretation of map symbols could be a matter of life and death. Military maps (see p. 183) and hydrographic charts (see p. 175) are highly complex, and explanations of their symbology are vital to their correct use. Such keys or *legends* can themselves be significant documents – *U.S. Chart No. 1*, jointly published by the National Oceanic and Atmospheric Administration and the Department of Defense's National Geospatial-Intelligence Agency, entitled *Symbols, Abbreviations and Terms used on Paper and Electronic Navigational Charts*, stretches to 132 pages, six of which are devoted solely to abbreviations (fig. 19). The equivalent publication in the UK (*Chart 5011*) is 73 pages long.

Not all maps demand such detailed explanations, but as maps became more complex, particularly through the nineteenth century, users needed more and more help in understanding them. During that time, different scientific phenomena were being mapped for the first time, and systematic series of detailed topographic maps were being created in many countries.

18 The legend of this American-produced military map, *Former Yugoslavia Series M709 1:50,000 Sheet Vareš*, explains the overprint showing the Bosnian ceasefire line as defined in the Dayton Agreement of 1995. Misunderstanding of such maps could lead to serious geopolitical issues.

No.	INT	Description	NOAA	NGA	Other NGA	ECDIS	
13		Underwater rock of unknown depth, dangerous to surface navigation					Dangerous underwater rock of uncertain depth
							Isolated danger of depth less than the safety contour
14.1		Underwater rock of known depth; inside the corresponding depth area	12 Rk	27 Rk / 21 / R		5	Underwater hazard with a depth of 20 meters or less
						25	Underwater hazard with depth greater than 20 meters
14.2		Underwater rock of known depth; outside the corresponding depth area, dangerous to surface navigation	5 Rk	4_5 Rk / 5 R			Isolated danger of depth less than the safety contour
15	35 R	Underwater rock of known depth, not dangerous to surface navigation	35Rk		35 R +(35)	10	Underwater hazard with a depth of 20 meters or less
						25	Underwater hazard with depth greater than 20 meters
16	Co Co	Coral reef which is always covered	+Co 3_1	Reef line			Dangerous underwater rock of uncertain depth Obstruction, depth not stated
							Isolated danger of depth less than the safety contour
							Safe clearance shoaler than safety contour
						12_8	Safe clearance deeper than safety contour
						25_6	Safe clearance deeper than 20 meters
17	5_9 Br 19/18	Breakers	Breakers	Br	West Breaker PA		Overfalls, tide rips; eddies; breakwaters as point, line, and area

19 An extract from *U.S. Chart No. 1*, published regularly in the USA by NOAA and the National Geospatial-Intelligence Agency. Critical for safety at sea, the publication explains the multitude of symbols and annotations on American hydrographic charts.

The features shown on such maps were not necessarily familiar to users; nor were the symbols used to represent them. As a result, the legend became established as a conventional feature on maps.

Cartographers generally work on the principle that the symbols they use should be recognizable, and if they aren't they should be explained in a legend. Even the earliest map-makers were aware of this – a map from Ancient Egypt in *c.* 1200 BCE showed the location of gold mines and carried explanatory inscriptions, including one that says: 'The hills from which the gold is brought are drawn in red on the plan.' Medieval maps often included similar descriptive legends. Text on Erhard Etzlaub's 1492 map of Nuremberg explained how to use the map and described the symbols he'd used – red dots for towns, for example. During the fifteenth and sixteenth centuries legends of this type became integral parts of maps by cartographers such as Christopher Saxton. They were commonly presented in highly ornamented cartouches or decorative panels (see p. 50). As map detail increased, so did the extent and complexity of their legends.

If there is insufficient space on a map itself, the legend may be presented on a separate sheet. The sheet accompanying Etzlaub's Romweg map ('the Way to Rome') of *c.* 1500 was a forerunner to what became known as *characteristic sheets*. The *Times Comprehensive Atlas of the World* comes with its own characteristic sheet which serves as a useful bookmark. A series of twenty-five maps of the Austrian Netherlands in 1777 by Joseph Comte de Ferraris was accompanied by a sheet entitled 'Explication' and this approach became particularly common as topographic map series were developed through the nineteenth century. They would be produced, by agencies such as the British Ordnance Survey, in the style of

20 As maps in the nineteenth century began to show new phenomena, users needed more explanation of what was shown. The legend of *Airey's Railway Map of Manchester & District*, 1880, explained the colours used to distinguish between train operating companies.

the series to which they referred, showing all symbols that may appear on any of the maps and were often as beautifully designed and engraved as the maps themselves. Instructions on the use and compilation of legends were even included in cartographic manuals of the time, notably those by Lacroix and Marie in 1811 and 1825 respectively – the latter being credited with the first use of blocks of colour in a legend to explain areas of colour on a map.

Whether to include all symbols on a legend is a decision for the cartographer based on the subject of the map and its expected audience. Some symbols may be designed so as to be self-explanatory, such as pictorial representations of towns and villages, as were common through the medieval period. But as soon as settlements are shown by dots of different sizes and colours depending on their population size and administrative status, a key becomes necessary. On thematic maps – maps depicting a specific subject rather than just the topography of an area (see pp. 151–63) – it is commonly assumed that the underlying base map, showing settlements and communications network, doesn't need to be explained, but that details of the main subject do.

Some maps demand a legend more than others – the multitude of colours on a geological map, subtle variations in colour and tone depicting statistical information such as infant mortality, voting patterns or wealth, as in Charles Booth's poverty map of 1898 (see fig. 77), or detailed icons representing various types of vegetation, could not easily be understood without a key. Similarly, where correct map interpretation is critical for the user's safety, a comprehensive, easy-to-understand legend will be essential.

21 *Norfolciae comitatus* by Christopher Saxton, 1579, from his atlas of maps of English and Welsh counties, includes a decorative legend explaining letters (typographic symbols) used to identify Norfolk's thirty-one administrative units or 'hundreds'.

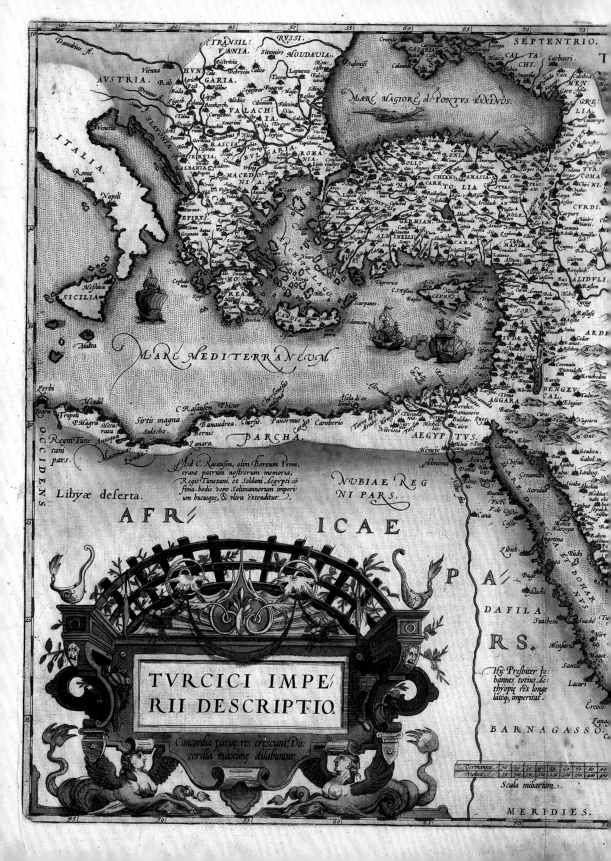

ORNAMENT
Art meets science

Chariots, castles, exotic and mythical creatures, coats of arms, political events and personal portraits: why might all these features appear on maps, when a map's purpose is to describe the lie of the land? Willem Janszoon Blaeu – of the Blaeu family, who were the leading map publishers during the seventeenth-century golden age of cartography in the Netherlands – published a beautiful world map in 1606 with a border featuring elaborate drawings representing the four seasons, the four elements, the planets and the Seven Wonders of the Ancient World (fig. 23). These can't be described as cartographic, but they certainly complement the map they surround and help to tell its story.

Embellishments have been common throughout the history of mapping. Maps were often used to promote a cause, to portray the power of an empire, country or company, or to hang on a wall as a work of art or as a statement of wealth. All cases gave the cartographer the opportunity to use ornament and illustration, reflecting the point at which art and science meet in the practice of cartography. Such artistic expression is found on map

22 Ortelius's 'atlas', published in seven different languages, contains over fifty maps, all remarkably accurate for their time and with a wide variety of cartouches and borders. Extract from map of the Middle East from *Theatrum orbis terrarum*, Abraham Ortelius, 1570.

23 OVERLEAF Beautiful cartouches and elaborate borders were characteristic of maps of the sixteenth and seventeenth centuries by the Blaeu dynasty and other cartographers in the Dutch Republic. *Nova totius terrarum orbis*, Willem Janszoon Blaeu, c.1606.

covers and to some extent within the design of maps themselves, but the most striking expressions of ornamentation on maps are in the form of *cartouches* and *borders*.

Cartographic cartouches are decorative emblems which commonly contain information about the map – its creator, purpose, title and date – but which can also carry elaborate illustrations relating to the map's content. They often show political influences or personal agendas behind the map, reminding us that many maps are symbols of power.

The oldest surviving map containing what could be described as a cartouche is a map of the world from 1448 by Giovanni Leardo of Venice. The heyday of such orna-mentation, though, was through the sixteenth and seventeenth centuries – a time when the Low Countries dominated the world of

24 *A Map of Yorkshire*, Estra Clark, 1949. This pictorial map is made up almost entirely of ornament and illustration. It was published to promote rail travel in the county, showing places of interest and details of the area's industry.

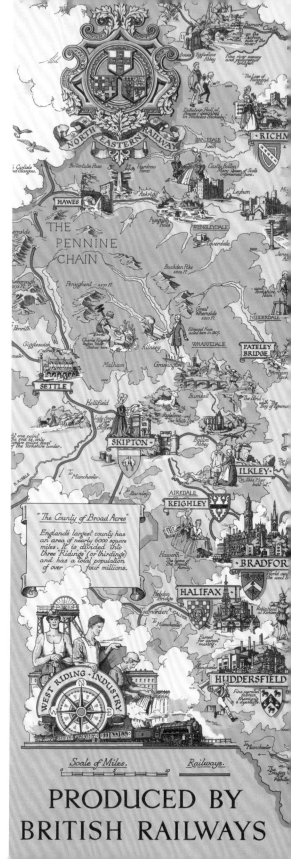

A MAP OF
YORKSHIRE

map-making. Designs developed through this period, reflecting the artistic styles and developments of the time – from relatively simple strapwork to much more elaborate late-Renaissance and Flemish baroque styles towards the end of the sixteenth century. The evolution of such ornamentation also reflected improvements in engraving techniques and in the way maps were reproduced, both allowing much more detailed work to be carried out.

Cartographers of this period – such as Ortelius, Mercator, Hondius, Blaeu and van Keulen – became masters of ornamentation. Abraham Ortelius's *Theatrum orbis terrarum*, published in Antwerp in 1570, is a particularly fine example of a highly ornamented publication, with each map supplemented by elaborate cartouches (fig. 22). This collection of maps was a highly influential publication in terms of both its content and its design – it is often referred to as the world's first 'atlas'. Ornamentation subsequently developed into more figurative artwork commonly depicting figures, cherubs, animals and pastoral scenes. Later examples carried symbolic images designed to portray the power and influence of either the cartographer himself or his benefactor – coats of arms, national flags on illustrations of ships, mythological figures representing certain virtues or professions, and so on.

Some cartographers copied and adapted artwork from 'ornament prints' or 'pattern books' which were published for artists and craftsmen – the forerunners, perhaps, to modern graphic swatches or clip art. Map-makers even adapted drawings of figures to represent leading personalities of the time. For illustrations on his 1650 map of France, Claes Janszoon Visscher copied from an earlier 1615 work by the Swiss draughtsman Mattaeus Merian. To bring his map up to date, Visscher simply amended a portrait of Louis XIII to be a likeness of his successor Louis XIV.

Many maps from this golden age also display very elaborate borders. They contain such features as detailed drawings of people from different social classes, economic activities and topographic views of landscapes and cities within the area of the map, and beyond.

25 This cartouche takes a more figurative approach to ornamentation than earlier maps. It shows shepherds and fisherman – presumably characteristic of the area covered by the map. *Glottiana præfectura inferior* (map of Lower Clydesdale), J. Blaeu, c. 1654.

However, the convention of making a political or personal statement through ornamentation was relatively short-lived. By the late eighteenth century the fashion for cartouches and borders had largely passed, with a 'scientific' approach to maps becoming more prominent. Titles, scale and descriptions still needed to be included, but they began to adopt the much simpler, functional styles with which we are familiar today.

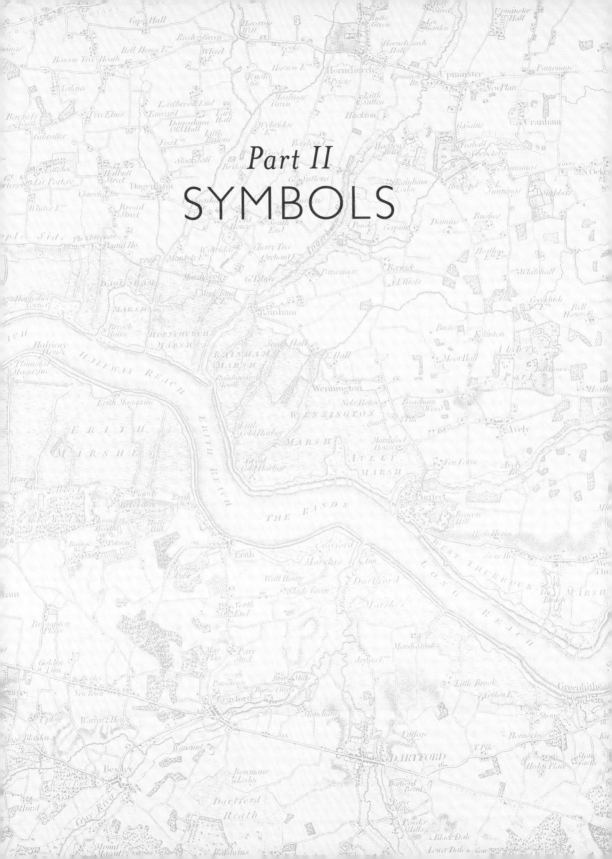

Part II
SYMBOLS

MAP SYMBOLS
Conventional signs

In May 2015, Ordnance Survey, Great Britain's national mapping agency, launched a competition with the BBC's *The One Show* inviting non-specialists to design some new symbols for its maps. It needed new symbols to represent such things as solar farms, electric-car charging points and skate parks. The audience rose to the challenge and sent in over 7,000 entries The winning designs are now finding their way onto the agency's maps (fig. 29).

Maps are obviously restricted in the amount of information they can show – partly because of scale (see p. 35), but also because of the complexity of the features and concepts they portray. In aiming to simplify and codify information to allow it to be mapped, symbols have become an integral part of what is sometimes described as the 'language' of maps. They come in three basic forms – points, lines and areas. Early cartographers had free rein in devising ways to symbolize features, and each was free to take a different approach. But more recently the use of certain symbols has become common practice, to such an extent that symbols are often referred to generally as 'conventional signs'.

Symbols fall into two basic styles. Pictorial or *pictographic* symbols directly resemble the features they represent; iconic or *ideographic* symbols

26 *Los Angeles SEC 103*, Federal Aviation Administration, 2018. Charts such as this are designed for flying under visual flight rules (VFR) and use variations in form, size and colour of symbols to ensure easy interpretation when a lot is at stake.

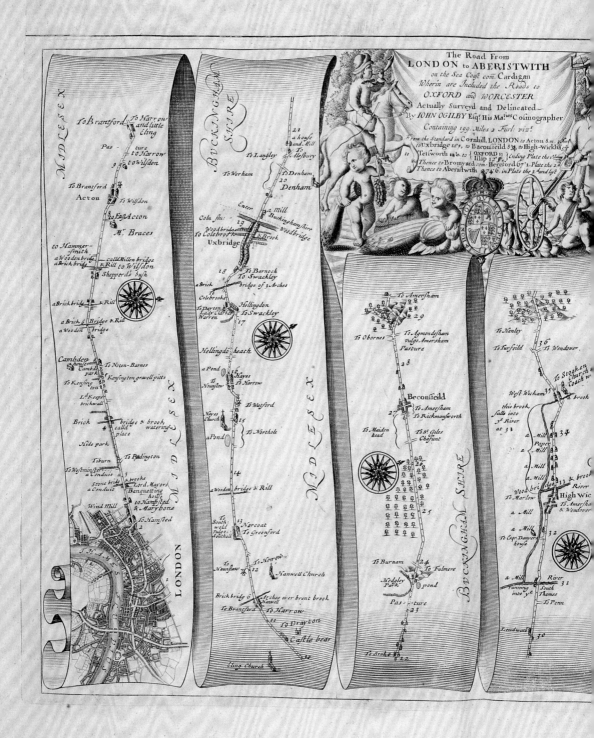

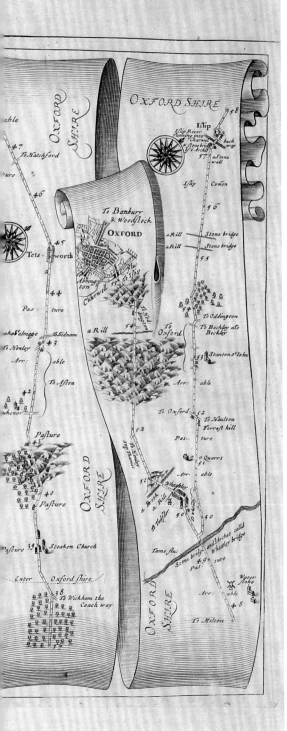

use geometric or diagrammatic signs to portray the idea or concept behind the features they show. Both styles have been used throughout cartographic history, but there has been a general trend from pictorial to more iconic representations.

A carved map from Nuzi, Mesopotamia, in the third millennium BCE depicts mountains in profile, and rivers as lines – early conventions which have lasted for centuries; and geometric symbols representing settlements and fields are evident on rock carvings from Brescia, Italy, dated *c.* 1500 BCE. A Roman map from the first century uses the now familiar technique of showing roads by double lines, and the later Roman map known as the *Peutinger Table* (fig. 28), while using single red lines to show routes, carries beautiful illustrations of villas and temples. This mixed approach continued with medieval

27 *The Road From London to Aberistwith,* from *Britannia* by John Ogilby, 1675. An excellent combination of iconic and pictorial symbols. Ogilby uses point symbols to mark specific locations and distances, double lines for roads and pictorial tree patterns for wooded areas.

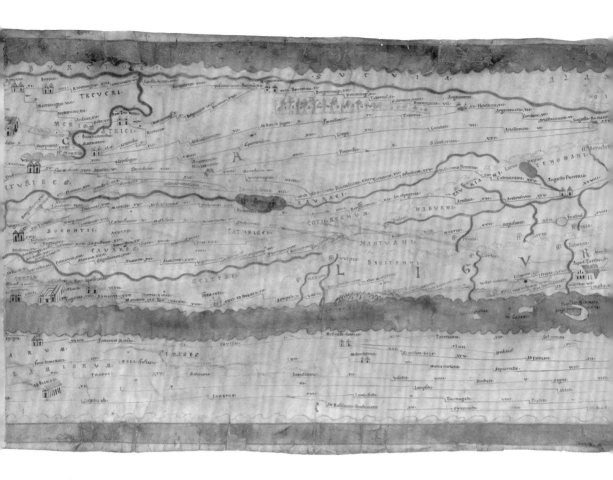

maps, although in general maps from this period were more pictorial than schematic. However, nautical charts from that time by Lucas Janszoon Waghenaer (1533–1606) used specific symbols for buoys and navigational beacons (see p. 175). Cartographers also began to classify towns by size and importance using pictorial symbols for larger settlements and conventional symbols for smaller places, such as the dots within circles used by Timothy Pont on his maps of Scotland in the late sixteenth century.

However, when national series of detailed topographic maps began to appear, especially in the nineteenth century, consistency between maps and a greater range of symbols were required. This coincided with the

28 The *Tabula Peutingeriana* (*Peutinger Table*), *c.*300 CE/*c.*1200. This twelfth- or thirteenth-century copy of the lost original shows its beautiful mix of distinctive red lines for roads (complete with distances) and pictorial symbols for villas, settlements and other landmarks.

29 Winning entries from the Ordnance Survey map symbol design competition, 2015. The challenge of designing symbols of this type is to make their meanings as obvious as possible, whilst making them sympathetic to the overall map style.

development and refinement of lithographic printing , which made it easier to print more complicated and consistent graphics. National mapping agencies devised detailed specifications for their maps, and whereas the meaning of pictorial symbols may be easily deduced, the more conventional ideographic symbols coming into use needed to be explained in legends (see p. 41).

The aim of symbolization is to ensure that features on a map are both detected and differentiated from one another. They also need to be clearly identifiable against a legend so that a user can extract meaning. To ensure these aims are met, cartographers use the *graphic* or *visual variables* – the characteristics of symbols that are adjustable to ensure their legibility. The most critical variables are form (shape), size and colour (hue and value or shade). By careful use of these variables a huge range of symbols exist – for example, point symbols of squares, circles and diamonds; dashed and double lines; areas filled with colour, pattern or texture. This – together with the classification of symbols into types of feature, such as water, settlements, veg-etation, roads – ensures that the map-maker's depiction will be understood.

There were some restrictions in the Ordnance Survey competition. Not all the variables were at the contestants' disposal. For example, only a single colour could be used and the symbols had to match the basic style of existing tourist symbols. But it nevertheless provided an insight into a challenge which has been familiar to map-makers through the ages – to create suitable designs that both look good and work well.

POINT SYMBOLS
X marks the spot

During a briefing in October 1962, US President John F. Kennedy drew
a series of small black crosses on a map of Cuba and added the annotation
'missile sites' (fig. 30). This marked the start of the Cuban Missile Crisis – a
critical period during the Cold War when nuclear conflict was a very real
possibility. US spy planes had captured photographs showing Soviet missile
sites being constructed in Cuba despite Premier Khrushchev's claims that
no offensive weapons would be deployed there. The crisis was resolved,
and the missiles removed, after a tense thirteen days.

Point symbols can take many forms to suit the purpose of a map. They
aim to locate features and to help map users identify them. Symbols may
be simple geometric shapes – squares, circles, triangles – or more complex
pictorial representations. Pictorial symbols, such as small perspective drawings of towns or buildings, give a clear impression of what the features they
represent look like in reality, whereas geometric forms bear no resemblance
and a key or legend (see p. 41) may be needed to explain their meaning.
Both types of symbol have their advantages and disadvantages. Geometric
symbols are generally easier to find on maps, although pictorial symbols
allow more accurate and immediate interpretation.

Preferences have changed through the centuries. The convention of
using circles to represent settlements dates back to Islamic maps of the
tenth century. Maps in the fifteenth century based on the work of Claudius
Ptolemy (*c.* 100–170 CE) use circled dots; maps by Erhard Etzlaub include

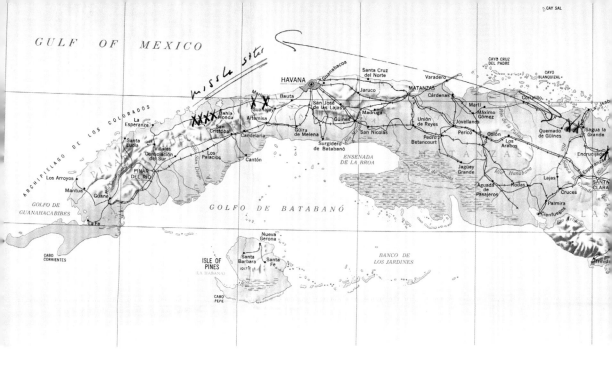

30 Detail from *Cuba*, CIA, 1960, annotations by President John F. Kennedy, 1962. Cartographic design and the choice of map symbols may take a back seat in times of crisis. The scribbled crosses marking missile sites immediately suggest the urgency of the situation.

31 OVERLEAF Extract from *The Gough Map*, c.1360. Over 600 settlements are shown on this unique medieval map of Great Britain, and clever variations of the pictorial point symbols used to depict them indicate their relative size and importance.

different sizes of symbol to indicate relative sizes and importance of towns. During medieval times, pictorial symbols came into fashion and persisted through the fifteenth and sixteenth centuries, sometimes taking the form of elaborate vignettes of towns and cities, often used in combination with circles. The drawings in some cases captured a likeness of each town; in others they would be standardized. Some degree of classification of features was possible, even with such elaborate representations – the map of Britain known as the *Gough Map*, from 1360, distinguished six types of settlement and also identified cathedral cities and monasteries by variations in the drawings (fig. 31).

A more 'scientific' approach to mapping was adopted in the eighteenth and nineteenth centuries as national topographic map series began to

appear, featuring standardized, geometric symbols. Both general maps and thematic maps (see pp. 151–63) forced cartographers to devise clear, simple point symbols to portray not only land features but also statistical information associated with those features or locations. Map-makers do this by exploiting the graphic or visual variables (see p. 61). Of these, form (or shape) and size are particularly relevant to point symbols and allow hierarchies of information to be shown – relative sizes of towns and cities shown by progressively larger circles, administrative centres shown as squares rather than circles, populations of settlements shown by circles drawn in direct proportion to their census statistics.

Abstract point symbols may now be dominant, but conventions are not universal and styles vary between countries, publishers and

32 This map uses simple graphic symbols (black open circles, some with crosses) for settlements – a style still in common use. More important towns are shown by pictorial symbols highlighted in red. Extract from *Map of Elgin and Moray, Scotland*, J. Blaeu, c. 1654.

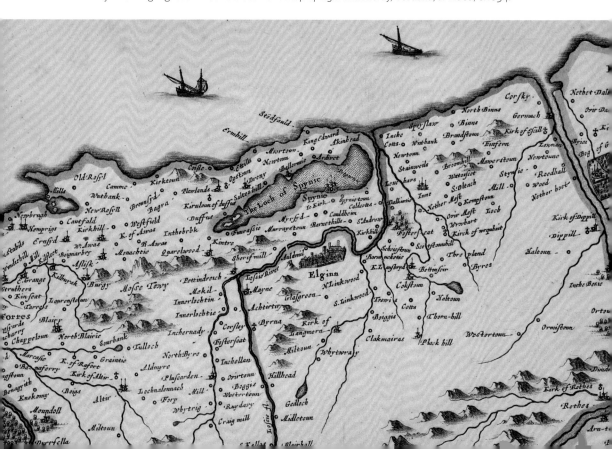

individual cartographers. Also, the split between geometric and pictorial symbols is not quite so simple. There is more of a continuum between the two, and the pictorial approach has certainly not disappeared completely. Pictographs lie somewhere in between. These are stylized graphic symbols which display key characteristics of the features they represent – simple drawings of ships to show ferry ports, buses to indicate bus stops, picnic benches for picnic sites. The British Ordnance Survey uses these to represent such things as television masts, wind turbines and windmills. This approach is common on tourist maps and many such symbols have become universally established conventions through their adoption and promotion by the International Organization for Standardization (ISO).

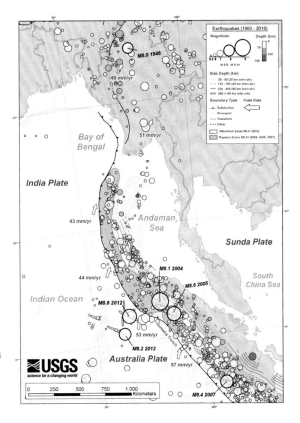

33 Careful use of point symbols on thematic maps can portray complex information in a clearly understandable way. Here, proportional circles of different colours help tell the story of this earthquake-prone region. *Map of Tectonic Summary Region – Sumatra*, USGS, 2016.

President Kennedy didn't worry about categorizing the Soviet missiles, or about what they actually looked like – irrespective of their type, each one presented a threat to his country. He just needed to locate them and to focus minds on the danger they presented. It was too urgent an issue to experiment with styles, so he used a simple point symbol – a device which has long been a staple tool of cartographers.

MAP OF THE
LONDON AND NORTH WESTERN RAILWAY AND CONNECTIONS.

SCALE 31½ MILES TO AN INCH.

LONDON & NORTH WESTERN RAILWAY AND ITS BRANCHES
LINES WORKED IN DIRECT & CONTINUOUS COMMUNICATIONS WITH DO.
LINES WITH WHICH THROUGH BOOKING ARRANGEMENTS ARE MADE L. N. W. RY.
OTHER LINES

— TEN DAYS' TICKETS —
Issued at GLASGOW, LIVERPOOL, LONDONDERRY and QUEENSTOWN.

Passengers from America who have not bought their Tickets in New York, can obtain at Glasgow, Liverpool, Londonderry and Queenstown Single Tickets for London, good for 10 DAYS, and giving the same Stop-over Privileges as Tickets bought in New York.
ASK FOR TEN DAYS' TICKET TO LONDON, EUSTON STATION.

THE IRISH MAIL ROUTE

ATLANTIC OCEAN

New York to Queenstown 2796 Miles

LINE SYMBOLS
Keeping on track

What better way to promote your railway company than to publish a map of your routes? The London and North Western Railway (L&NWR) did this in 1898 (fig. 34). Their map of England, Wales and southern Scotland prominently shows, in bold red lines, their own routes and those of partner companies. It makes it clear that they have Britain covered and to get anywhere you will have to use their trains. But look a bit closer, and you see that there are other lines on the map. Apologetically thin black lines are used to show the routes operated by rival companies, including the important east coast route between Edinburgh and Newcastle. It seems the L&NWR was not the only option for travelling around the country after all.

This map makes good use of the graphic or visual variables used by map-makers to design symbols that represent features in ways that make them distinctive. The L&NWR routes are made prominent by using the variables of form (solid lines), size (the lines are thick and unmissable) and colour (using prominent, advancing, 'important' red – see p. 88). In contrast, the poor rival routes are very thin lines in neutral, unremarkable black. Clever use of these variables with line symbols gives great scope for cartographers to distinguish between the multitude of linear features

34 *Map of the London and North Western Railway*, 1898. In a style typical of the time, this map of L&NWR routes gives them clear prominence over competitors' routes by the use of colour and by varying the thickness of lines.

they may need to include on their maps. A general or topographic map, for example, may include dashed lines for disputed or undemarcated boundaries (see p. 137), dotted for walking routes; solid blue for permanent rivers, pecked blue for seasonal streams; double lines for main roads, some with a fine centre line for dual carriageways; brown contours across farmland, black across rocky terrain (see p. 111). On thematic maps, too (see pp. 151–63), the thickness of lines can be proportional to statistical values (imports and exports between two countries, for example), or red and blue arrows can indicate warm or cold airflow on weather maps.

Maps have always been based on line symbols. It isn't surprising that lines representing rivers appear on ancient maps of Mesopotamia – a region whose name means 'land between rivers'; nor that Roman maps commonly carried straight lines to indicate direct routes between settlements. During the medieval period, boundaries of countries and counties began to appear regularly on maps, commonly as fine broken lines enhanced by hand-coloured vignettes. But movement has been a prominent feature in the use of line symbols – rivers flow, people and goods move along defined routes, ships cross oceans. Maps of China from the first century BCE indicate the importance of rivers to travel at that time, and the famous stone-engraved map the *Yu ji tu*, from the twelfth century (see fig. 13), provides a remarkably accurate picture of China's great rivers.

Developments in cartographic techniques through the nineteenth and twentieth centuries, and in road and rail travel, provided great opportunities for cartographers to fully exploit variations in line symbols. The golden

35 Extract from *Book on Navigation* by Piri Reis, 1525. A conventional blue line representing the River Nile dominates this map. Compass directions are shown by different coloured lines and neighbouring mountains are portrayed using an elaborate pictorial line symbol.

36 OVERLEAF *North Atlantic Ocean* by J.G. Bartholomew, 1907, uses different line symbols to show transatlantic shipping routes, bathymetric contours (isobaths), submarine cables, average limits of Arctic drift ice and the route of HMS *Challenger* during its expedition of 1872-76.

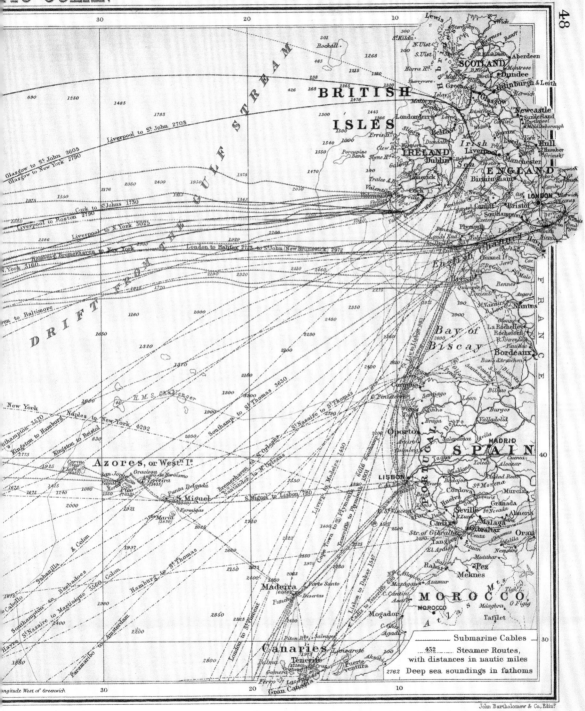

Submarine Cables
432   Steamer Routes,
with distances in nautic miles
2762 Deep sea soundings in fathoms

John Bartholomew & Co., Edin?

age of motoring between the 1930s and 1950s saw oil and tyre companies publishing huge quantities of road maps, usually distributed free. Careful use of colour and line form would distinguish just how motorable the different roads were – in some cases up to forty or fifty classifications of route could be identified.

Routes don't always need to be shown by accurately plotted lines. The sequence of settlements or stations may be equally or even more important than the precise alignment of the route. The classic example of this type of *topological* map – one showing just a network of lines and the points where they join and intersect – is the London Tube map, originally created by Harry Beck, which distinguishes between the lines primarily by the use of colour (fig. 37). Lines become the dominating feature of such maps – a technique the creators of the L&NWR map certainly understood and used to their advantage.

BREAKING THE CONVENTION

John Ogilby's *Britannia*, published in 1675 (see fig. 27), introduced the revolutionary idea of line or 'strip' maps following the routes of selected roads across Britain – 7,500 miles of roads in total. Arranged in vertical strips across the page, each map follows the line of a route in great detail, marking distances and showing whether sections of roads are enclosed by walls or hedges, or open to the countryside or cultivated land.

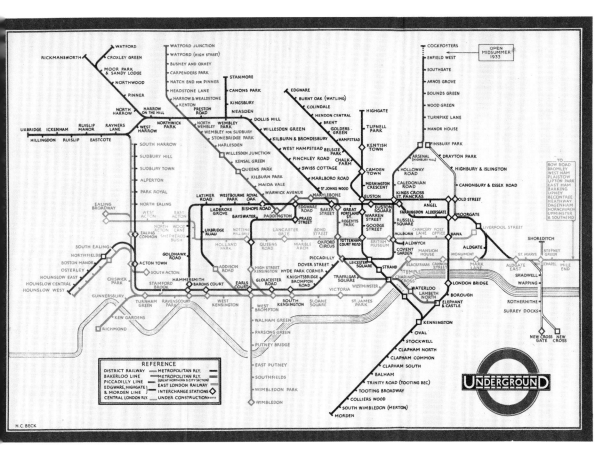

37 *London Underground* by Harry Beck, 1933. For navigating the tube, Beck judged true geography to be less important than being able to identify lines and their intersections. His careful alignment and colouring of routes has made his map a design classic.

AREA SYMBOLS
What's the difference?

In the middle of the First World War, Sir Mark Sykes (UK) and François Georges Picot (France) were given the job, by the British and French governments, of dividing up part of the Ottoman Empire. The Ottomans were allies of Germany, and Britain and France were keen to break up their powerful empire and share their land between them. After some negotiation and various (subsequently unfulfilled) promises of autonomy for Arab groups in the region, they took a Royal Geographical Society map of the Middle East and drew the 'Sykes–Picot line' across what is now Jordan, Syria and Iraq, identifying areas which Britain and France would govern or retain some influence over, and an area bordering the Mediterranean Sea which would be under international jurisdiction (fig. 40). Lines on the map weren't sufficient to depict what was agreed, so specific areas were coloured blue, red and yellow to clarify the situation.

Area symbols such as the colours used by Sykes and Picot are important in mapping discontinuity on the Earth's surface. In this case, new political breaks were created; in other examples there may be physical boundaries – cities in contrast to rural area, differences in rock type, forests adjacent to arable land – or social and political divisions which need to be mapped. Mapping such differences has been a part of cartography for centuries – the

38 Extract from *One-inch 'Old Series' Sheet 1. London*, Ordnance Survey, 1805. A fine example of the use of beautifully engraved pattern fills to distinguish buildings, marsh, cultivated land, woodland and the River Thames – a great cartographic challenge without the use of colour.

basic depiction of areas of land and sea by different colours is found on Chinese maps from the second century BCE. In an early political statement, a portolan chart by Bertran and Ripol in 1456 coloured the extent of the Moorish kingdom of Granada in green. This was a period when the idea of the nation-state was developing and maps became an important part of showing just who owned what. Similarly, estate owners in the sixteenth and seventeenth centuries went to great lengths to map the extent and nature of the land they owned.

Mapping physical differences or clearly established geopolitical boundaries (see p. 137) may be straightforward, but as science, surveying and statistics developed in the eighteenth and nineteenth centuries more complex phenomena, such as climatic, environmental and social conditions, needed to be mapped. More accurate land survey was allowing detailed contours to be plotted, and this in turn led to the first use of layer colours – an area symbol indicating land elevation – by Jean-Louis Dupain-Triel in 1798. Data from national censuses (which began in the late eighteenth century) related to defined areas, and cartographers had to find ways to make statistical differences and patterns clear. Thankfully, technology kept pace with this requirement. Previously, maps had been hand-coloured (as in the case of the Sykes–Picot map), which limited the range and consistency of colours that could be used. New lithographic printing techniques of the nineteenth and twentieth centuries, however, allowed subtle differences and consistencies in colour to be applied to areas.

Colour is not the only way to distinguish between areas, although it is the most common graphic or visual variable used. Because the shape and size of areas to be mapped are defined by geography, these variables cannot be used as they can with point or line symbols. However, areas may

39 Estate maps such as this were used to manage the collection of tithes. Different colours and patterns indicate land cover, such as pasture and arable land, as well as areas for which tithes were due. Extract from *A Survey of the open fields and old inclosures in ... Benson*, Richard Davis, 1788.

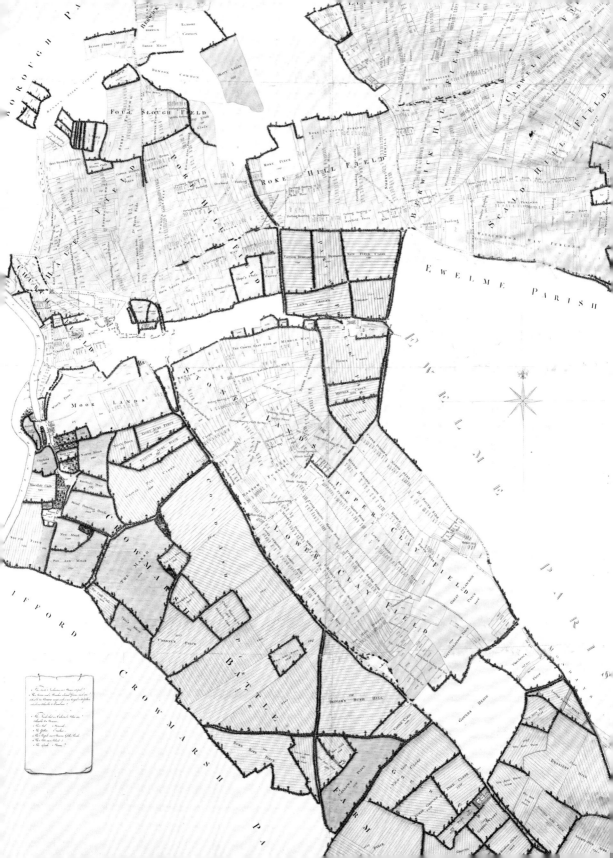

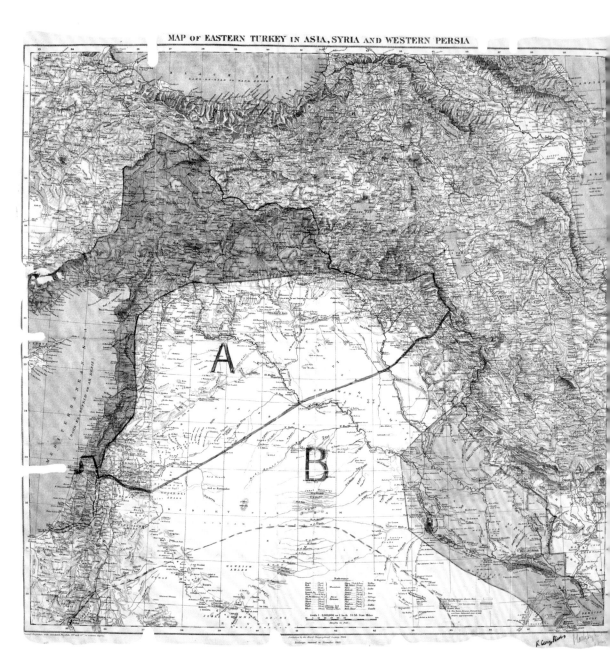

40 *Map of Eastern Turkey in Asia, Syria and Western Persia*, Royal Geographical Society, hand-coloured during negotiation of the Sykes–Picot agreement, 1916. Blue and red represent areas under direct French and British rule respectively, while yellow indicates the area under international administration.

be represented by patterns made up of individual point symbols or lines, and the size, spacing and orientation of these patterns – the repetition of tree symbols to represent woodland, for example – are variable. If then combined with variations in colour – for example, the conventional representation of built-up areas, popular in early topographic map series, using tints of red or black, with fine line patterns – a huge range of area symbols becomes available.

BREAKING THE CONVENTION

In the sport of orienteering, it is important to know where there are areas of woodland, but much more important to know how easy those areas are to run through. For this reason, orienteering maps depart from the convention of showing woodland as simply green. Open 'runnable' woodland is shown as white, while different shades of green indicate relative difficulty, with the darkest green showing impenetrable woodland or 'fight'.

As effective as area symbols may be, there is a danger that they will oversimplify things, and have a powerful influence on events or on a map user's way of thinking. Areas represented by uniform colour are not as homogeneous as they may look – important internal socio-economic differences can become conveniently 'invisible' – and ownership of territory is often contentious. By using red and blue on the Sykes–Picot map the British and French were clearly stating 'this is ours, we are in control', but the creation of the boundary and the colouring of the areas perhaps created differences and conflicts which were not previously there.

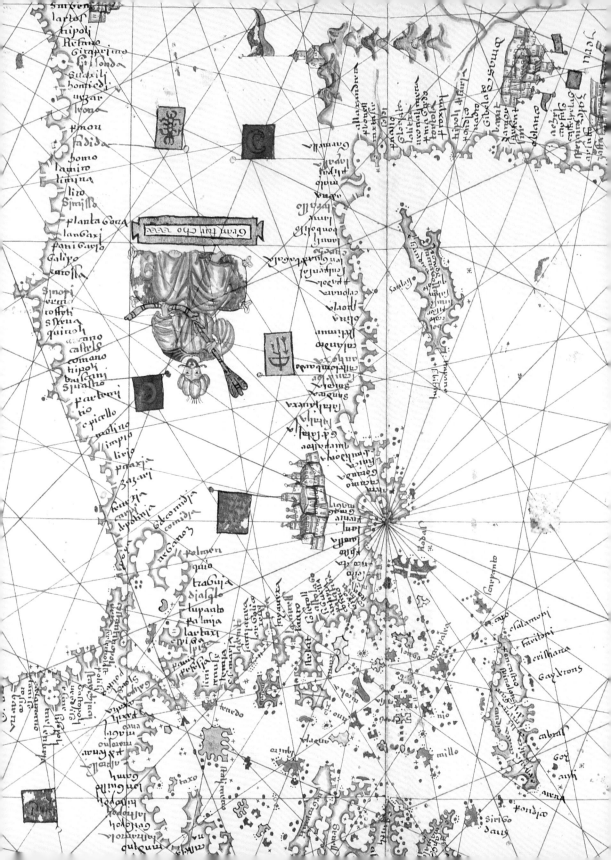

COLOUR

Deep blue sea?

First … colour over all the Hills … with the Tincture of Myrrh very thin; then if there be any Woods, dab every tree with the point of a very fine Pencil dipt in Grass Green … then with another Pencil dipt in Red Lead … let the Principal Cities and Towns be done over that the Eye may more readily perceive them.[3]

John Smith's *The Art of Painting with Oyl*, published in 1701, includes this detailed advice on hand-colouring maps, which was at that time a popular, genteel pastime. Map production had steadily developed from original manuscript drawings, through woodcut printing in the fifteenth century to copperplate engraving in the seventeenth century. This latest production technique allowed easier reproduction of more detailed maps, but the technology still wasn't suitable for the printing of more than one colour. Hence the practice of hand-colouring engraved maps. As well as being a forerunner to today's use of stress-relieving colouring books, hand-colouring became a specialized and respected occupation. Colour reproduction was made possible through the invention of lithographic printing in 1796 and subsequent refinements, and the routine practice

41 This extract from a portolan chart of the eastern Mediterranean Sea by Bartolomeo Oliva, 1559, makes extensive use of colour in its illustrative and ornamental features, but also uses colour to highlight more important settlements and to differentiate between compass and wind directions.

42 OVERLEAF *Typus orbis terrarum*, Abraham Ortelius, 1606. A beautiful example of a hand-coloured map, showing early use of the conventions of blue-green for the sea, red for towns and cities, and bands of colour for boundaries on land.

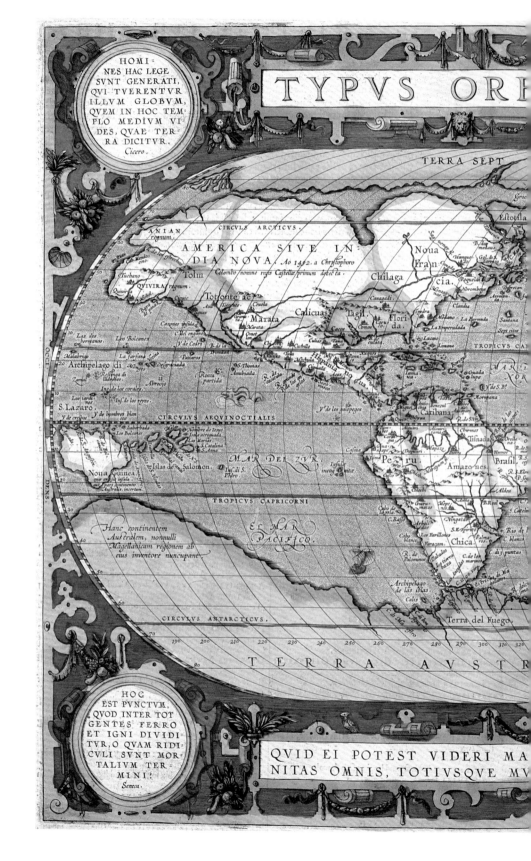

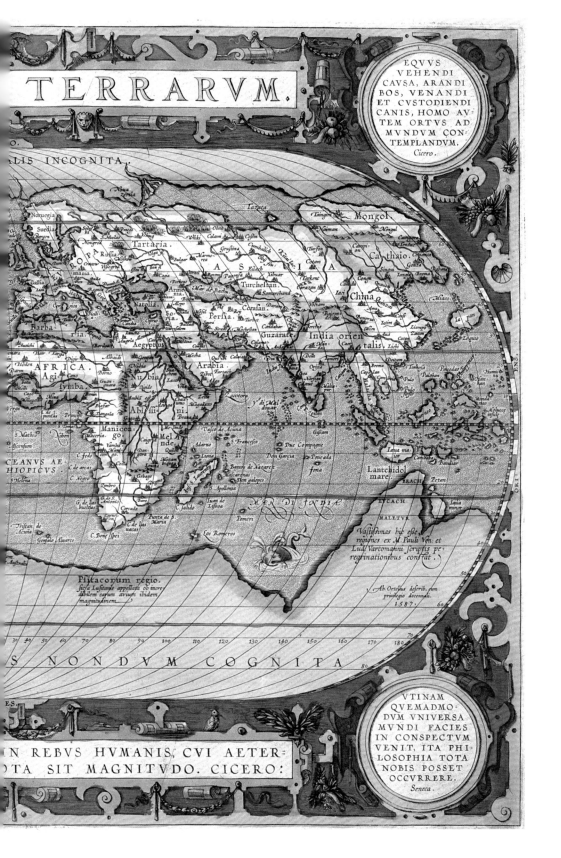

of hand-colouring of maps had disappeared by the end of the nineteenth century.

Colour has played a crucial role in mapping since ancient times. It is the most important 'graphic variable' (see p. 61) used by cartographers to distinguish between symbols and to emphasize important information. It also has emotive connotations which may be used by map-makers for good or bad – red signifies danger or power, green is soothing or natural, blue is cold and negative. Such associations were commonly used in wartime propaganda maps. Maps from Ancient Egypt and the Chinese Han dynasty (second century BCE) used colour to distinguish between features – Han maps showed information of military importance in red, water features in light blue-green, and other features and text in black. But wider conventions were not easily established. Different

43 *Queen Victoria's Empire*, W.S. Cater & Co., 1891. The use of red – an emotive and visually dominating colour – was widely used on maps of the British Empire to signify its importance and power.

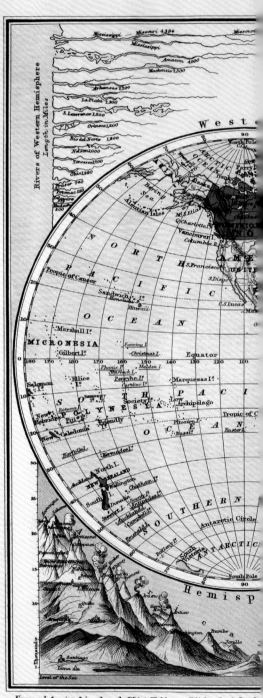

Engraved & printed in colours by W.& A.K.Johnston, Edinburgh & Londo[n]

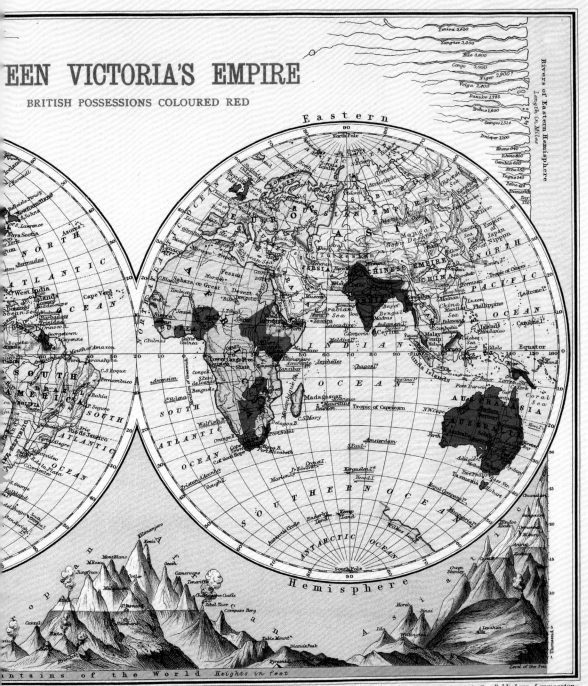

EEN VICTORIA'S EMPIRE

BRITISH POSSESSIONS COLOURED RED

cultures had different perceptions of colour, and its use depended on the knowledge and availability of the (sometimes rare and expensive) pigments required. Symbolism of colour in ancient times varied between cultures. The Babylonians, for example, associated certain colours with the cardinal points – black, white, red and yellow for north, south, east and west respectively – and the Greeks designated colours for the elements – white for air, black for earth, red for fire and ochre/yellow for water. Similar schemes related to the four seasons, and portolan charts of the thirteenth and fourteenth centuries showed wind directions by employing different colours.

The use of red for particularly important features dates back to some of the earliest mapping. This practice was established long before our current understanding of light, electromagnetic radiation and optics, and of colours being either 'advancing' (red) or 'receding' (green or blue). Medieval cartographers mastered the use of colour and produced some of the most beautiful hand-coloured maps. By then water features were more universally shown as blue or green, settlements as red and mountains as brown or black. During the Renaissance the idea developed of representing the natural world more accurately and depicting something of the character of the landscape. Early estate maps used colour and pattern to distinguish between cultivated and uncultivated land, and different shades of green suggested the richness of vegetation. General maps became more systematic in their use of colour, in particular in their depiction of boundaries, roads and water features.

Scientific and social changes in the later eighteenth century demanded new mapping techniques. Colour printing couldn't keep pace with the developments in, for example, geological mapping (see p. 167), which required accurate and consistent areas of clearly distinguishable, and increasingly conventional, colours. Statistical thematic mapping (see p. 157) needed steady progressions of colour to give indications of value, such as blue to red progressions for the range of average annual temperatures.

Current colour conventions – blue water, green woodland, brown terrain, red or black settlements, red roads, black railways – had become well established by the mid-nineteenth century.

Developments in printing and in digital mapping now give cartographers a great degree of control in the design of their maps and the use of colour. Do-it-yourself cartography still exists but has moved on from hand-coloured printed maps. Online mapping applications now allow users to choose the data and features to be mapped, as well as their colour and style. It took centuries for colour conventions to become established; perhaps new mapping techniques will both demand and generate new conventions to serve modern styles and methods of mapping.

S C O C I A
WALLIA
Catenes
Galeweia
Rokesburc
Beverlac
Stanford
Dunestapl
London

GENERALIZATION
Fake maps?

Matthew Paris (*c.* 1200–1259) was a Benedictine monk at St Albans Abbey who worked as an illuminator, chronicler and cartographer and created some of the world's most beautiful medieval maps. His best-known maps are decorated strip maps, or *itineraries*, detailing pilgrimage routes, in particularly the route from London to Jerusalem. On a more general map of Britain (fig. 44) he faced a problem. A note on the map says that 'If the page allowed this whole island should be longer.' His struggle to fit all his information onto the page in front of him is one common to map-makers throughout history.

In compiling any map, a cartographer needs to have in mind its purpose, scale and – Paris's downfall – its intended size. Map-makers must make compromises, particularly for small-scale maps, to ensure a map serves its purpose. As scale decreases, it becomes more difficult to show what you want, in a way you want and to the level of accuracy you might prefer. To tackle these issues, to ensure that the map is legible and in an appropriate style for its scale, cartographers carry out 'cartographic generalization'.

The level of information shown on a map needs to be appropriate, and so some information available to the cartographer will be judged irrelevant and omitted, and some will be classified into distinct categories. Scale

44 Map of Britain by Matthew Paris, from *Historia Anglorum*, c. 1250–59. Fitting the whole of Britain onto the page appears to have been a problem. Paris appears to have adjusted the shape of the land to ensure it could all be shown.

imposes limits even on this fundamental process. On a relatively large-scale regional map, roads may be described as motorways, primary roads, A roads, B roads, minor roads and tracks. But for the same road network to be shown sensibly on a small-scale national or continental map the classification has to be simplified to perhaps just motorways, main roads and other roads. The 'accuracy' is reduced, but the map still fulfils its main function. This applies also to thematic maps (see pp. 151–63) where statistical classifications need to be appropriate to both the data and map scale.

Once the information has been compiled, the issue of symbology arises (see p. 57), and the key elements of generalization come into play – combination, exaggeration, displacement, omission and simplification. Individual buildings are combined into general built-up areas, and adjacent areas of woodland are joined together. On all but the largest-scale maps, road widths are hugely exaggerated on maps compared to their true size (if they were drawn to their true scale the lines would be so thin they would be barely discernible). Roads may be widened even further, and features displaced so as to be kept on the correct side. Some detail is omitted, while complicated drainage patterns, and courses of individual meandering rivers, are simplified. There are no conventions or rules to these processes, which has made them very difficult to automate in digital mapping systems; rather, they are based on the skill and knowledge of the cartographer and some general principles. As long as the purpose of the map is kept strictly in mind, the character of the area mapped is maintained and the treatment is consistent across the map, the process will be successful (for example, lakes in Finland (fig. 45) could not be excluded from a map purely on the basis of size – some will need to be retained to indicate the generally 'wet' nature of the region).

45 Extract of map of Scandinavia from the *Times Atlas of the World*, Mid-Century Edition, John Bartholomew & Son, 1955. Not all of Finland's lakes are visible at this scale, but careful generalization ensures a good impression of the landscape.

A TAPESTRY OF TIME AND TERRAIN
By
José F. Vigil, Richard J. Pike, and David G. Howell

But without clear rules, is there not a danger of things being taken too far? Are there limits to how much exaggeration, displacement or omission can take place? Arrow symbols, circles, colour, map projections and place names are among what have been described as 'cartographic assault weapons',[4] and may take the principles of generalization to extremes. Maps have long been an important tool in propaganda, relating to issues such as territorial disputes, war, inequality and perceived threats. Projections (see p. 22) can exaggerate area, arrows can be large and bold, neighbouring countries can be mapped in 'threatening' colours (see p. 83) and place

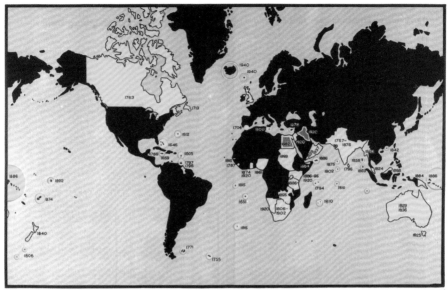

Englands Raubzug über 5 Erdteile 1605-1940

46 *A Tapestry of Time and Terrain*, USGS, 2000. This beautiful, complex geological map pushes generalization to the limit. Classification of geological time units into distinct categories helps interpretation, but the map still needed to be published with an explanatory booklet.

47 *England's Raids over 5 Continents 1605-1940*, Alois Moser, 1941. A wartime example of very selective information, portrayed dramatically, taking cartographic generalization – particularly selection, simplification and exaggeration – to extremes for propaganda purposes.

names translated into different languages in order to argue a point of view (see p. 131). The Nazis in particular recognized this and between 1933 and 1945 used maps as propaganda to promote their cause.

Matthew Paris's itineraries served more practical and spiritual purposes – allowing 'pilgrims' to travel on imaginary journeys to significant Christian sites. Like all cartographers, by being clear about the purpose of his maps he employed generalization, probably without realizing it, to make his maps more usable.

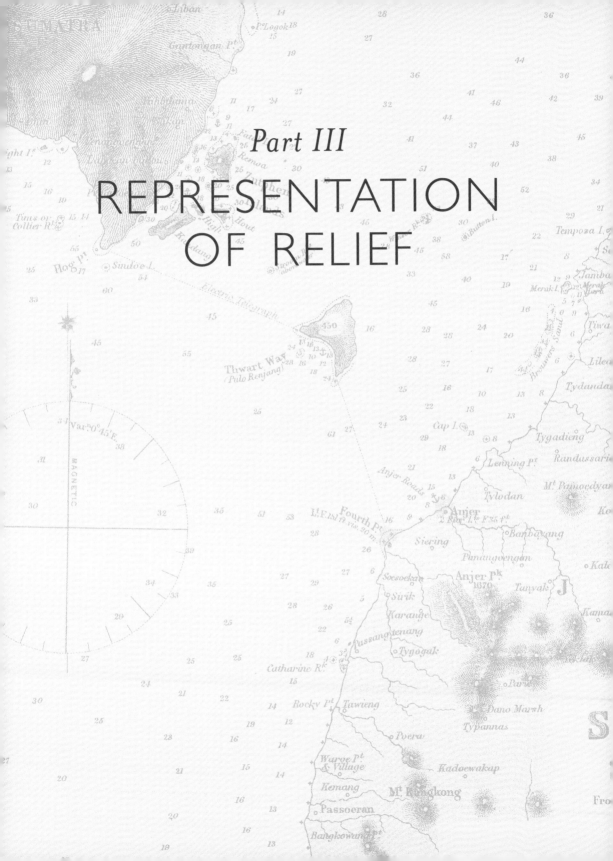

Part III

REPRESENTATION OF RELIEF

Ilft. Sneeck. Balck. Doeut faut.

Smal-brug. Myrnis. Schten houck. Zuyder gatt. 3

Oldega. Nyega. Hemelum. Hille gatt. 2½ 4 3 Kete gatt. 3¼

Abbega. Olde ga. 2½ 2 2½ 2 2 2 1½ T.

Hefacle. Coldu. Warns. De Kil. Enchuſen. Bucaſpel.

waert. Groutorp. Scharlc. Croepel ſant. Groote brouck. Veenhuſen. Hē.

Exmora. Molewerum. Staueren. Wage patt. Lutke brouck. Binnewijs.

Pandegra. Hinlopen. 4 Hooth car.

Voorwolde. worcum. D. Kil. Brouch oort. Weſt n.

Albnga. Peangum. Kreil. 5 De ſtraet 3 Sn.

Maccum. Idſhuyſen. De bocht. Plaet. De ſtraet Wernerſhoef. We

rroert. Weſt va Worcum. 2½ 2½ 3 5 Memelick. Ooſtwout.

ruer ba. Middel gront. 3 t bree ſant 4 3 Opperdues. Almerdorp. Sp

hen. 4 5 4 De gamels. Twiſch.

6 0 Breeſandt. 3

6 Middel gront. 2½ tflack Ertſwo

Sluytefant 2½ 2¼ Sunghorn

6 2 3 Winckel

t oude vlie Doue balch 5 Wieringe Doeuer Colhorn

Schierlings hals. De Nes. 8 Vlieter Barn hor

Drieſe Ree. 10 De Nes. 2 Repel 3 4 Repel

ke ſloot Nes. Ooegel ſant. 4 Harin

10 5 Schag

11 12 De boß Balch 5 De Waert Keyn

Robbe ſant. 6 13 5 Tnieuwe diep De

lant Weſton 3 De Pan 15 4 mars diep 12

2½ Ooſten 8 Geeſt niet 12

2½ Teyerlant. Gaeſt 12 Hunſduynen.

3 D wael Borch Geeſt

5 Texel. Horn. 15

8 3 4

SPOT HEIGHTS & SOUNDINGS

On the dot

Common sayings often have less well known and unrelated origins. Before an important committee meeting, the chairman may 'take soundings' by asking people how things to be discussed might be received. A poor decision may leave the committee accused of having 'plumbed the depths' by going for the worst option. Both phrases have nautical origins, which are part of the story behind the representation of height and depth on maps.

For safe navigation at sea it is critical to know the depth of water. For centuries this was measured by dropping a lead-weighted line – a *plumb* – overboard and noting when it hit the bottom. The measurement taken was a 'sounding'. Modern soundings, which are key components of hydrographic charts (see p. 175), are taken (appropriately for the name) by sonar or echo sounding – a distance measurement based on the time taken for sound waves to be emitted and reflected off the seabed.

Sea depths first appeared on a chart of part of the Flanders coast made by Pierre Pourbus in 1551. Early charts often had descriptions of depths rather than point symbols, but the first English chart to use numbers was of the River Humber in 1569. Dutch draughtsmen mastered the art of mapping coastal soundings, notably in Lucas Janszoon Waghenaer's beautiful 1585 atlas of sea charts, *Spieghel der Zeevaerdt* (fig. 48).

48 Waghenaer's atlas of sea charts was revolutionary in its depiction of depth soundings – critical information for safe navigation (see also fig. 88). Extract of chart of the Zuider Zee from *Spieghel der Zeevaerdt*, Lucas Janszoon Waghenaer, 1584.

The representation of height on land lagged behind soundings because of the difficulties of accurate measurement. Barometric pressure was one early means of measuring height, following the discovery by the French scientist Blaise Pascal that pressure decreased with altitude. Christopher Packe's chart of eastern Kent in 1743 showed heights converted from barometric readings. More accurate land surveying methods developed rapidly from the late eighteenth century and allowed spot heights to be plotted more precisely.

Packe's figures referred to altitude above a point he had selected at Sandwich, Kent. Spot heights and soundings need a fixed point against which both land heights and sea depths are to be measured, and it is vital for a map or chart to state just which 'vertical datum' is used. Some early-eighteenth-century maps measured land heights downwards from high points in negative values, but the fixing of zero at sea level was first used on a map of Midlothian, Scotland, by John Laurie in 1763 and became fairly universal by the end of that century. The importance of accuracy in establishing a datum is illustrated by the fact that the British Ordnance Survey's datum of mean sea level at Newlyn, Cornwall, was established from measurements taken every hour between 1 May 1915 and 30 April 1921. Each country has its own defined datum on which its mapped height data is based. France, for example, uses sea level at Marseille; Spain at Alicante.

Spot heights on land and soundings at sea are represented by point symbols (see p. 62) – most commonly simple dots, but also diagonal crosses, circles or even anchor symbols – accompanied by the measured value. Units of measure are most commonly metres, feet or – for depths – fathoms (6 feet). Placement and density of points are at the discretion of cartographers, but are dependent upon how many surveyed points they have to choose from, and will depend on the scale and purpose of a map or chart. Spot heights provide elevation information but don't, in themselves, give a detailed impression of the overall terrain. When combined with contours, though, they become useful in helping the user interpret the lie of

49 The survey behind this highly detailed chart involved taking over 13,000 depth soundings. The regular lines of soundings show the routes taken by the survey vessels. Extract from *A Map of the Extremity of Cape Cod*, US Bureau of Topographical Engineers, 1836.

the land. Careful placement will help in this – summits, cols, depressions, river confluences, important urban landmarks, road junctions, lakes are all good places for spot heights. Soundings are not intended to interpret the overall shape of the seabed, but are there primarily to provide critical information to seafarers, who only need to know a safe route. Deepest points, shallows, sand banks, entrances to navigable channels and the edges of continental shelves are the places they will expect to find this information.

Modern survey methods generate huge amounts of elevation data. Continual soundings taken over vast areas and aerial surveys measuring dense networks of points for the creation of digital terrain models allow dramatic visualizations of the seabed and landscapes. It would be 'stretching the point', though, to suggest that these have replaced the need for accurate treatment of simple heights and depths for safe navigation and the interpretation of terrain.

BREAKING THE CONVENTION

Soundings commonly appear on charts in straight lines reflecting the routes taken by the boats carrying out the depth surveys. A chart of Squam Lake in New Hampshire, USA, produced by celebrated cartographer Bradford Washburn in 1968, however, shows soundings in a beautifully neat, regular grid pattern. His survey was in winter and he chose to take precisely spaced soundings from the lake's frozen surface.

50 Spot heights may be more significant in flatter areas, where contours or hachures are scarce. With careful placement, they give at least an indication that the land is not completely flat. Extract from *Keswick (Cockermouth), One-inch Series, Sheet 9*, Ordnance Survey, 1897.

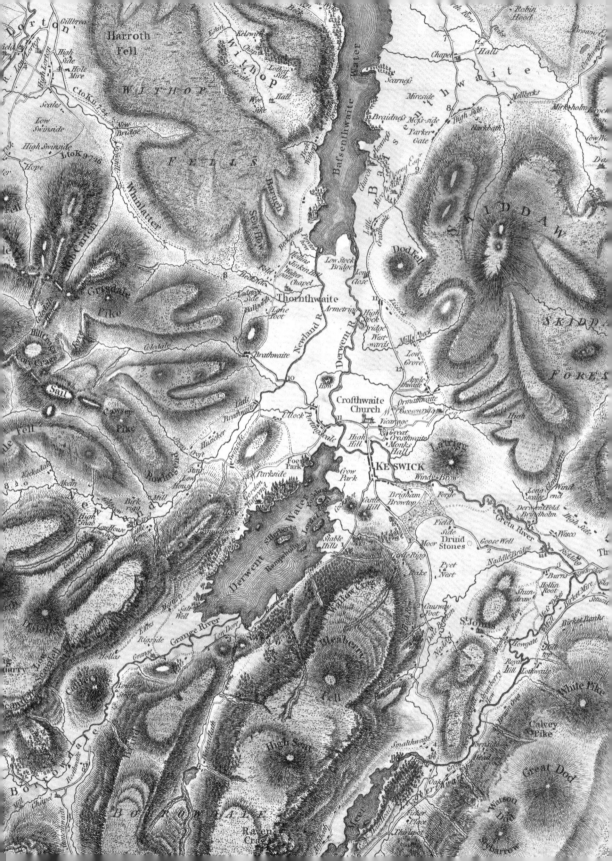

HACHURES
Sloping off

When a hillwalker needs to find a route up a mountain, or a child enjoying the snow with her sledge wants to go even faster, the actual height of the land around them is not important. What is important is the relative steepness of the slopes they face – the hiker may want to avoid the steepest way up, the daredevil child may prefer the steepest way down. The easy, perhaps innate, ability to estimate slope angles, without the need to know precise heights above sea level, lies behind the use of *hachures* to portray relief on maps.

Hachuring techniques are closely related to those of hatching used in the graphic arts – the creation of patterns of fine lines, often parallel, to indicate distinctions between types of surface. These artistic methods have been used to portray hills and mountains in pictorial form, often in elevation (side on) view, on maps since ancient times. Hachures are a development of these techniques for use on maps in plan view. They consist of lines running downslope in the direction of the steepest slope. Angles of slope were originally estimated by observation – they developed before techniques for measuring land height were developed. Today, contours are a valuable tool for interpreting slopes and for creating more accurate hachures.

51 Hachures indicate slopes, and an early form of layer colouring gives an indication of relative height, in this dramatic portrayal of the mountainous landscape of the Lake District. Extract from *An Actual Topographic Survey of the Environs of Keswick*, William Faden, 1789.

The Maltese cartographer Giovanni Francesco Abela was the first to use hachures on his map of the islands of Malta in 1647, while in France David Vivier used them on a map of the Paris area in 1674. Christopher Packe's 1743 map of eastern Kent depicted the general lie of the land – valleys and hills – with hachures, and was also the first map to use spot heights to show true elevations. The first use of hachures on a world map was in 1752 by Philippe Gouache. Techniques were steadily refined, and reflected technological developments in mapping, particularly that of copper engraving, which began to allow precise control of the thickness of lines.

52 An example of the *Dufourkarte* of Switzerland, illustrating the exemplary use of engraved shadow hachures to indicate slopes and to give a three-dimensional impression of the alpine mountains. Extract from *Topographische karte der Schweiz*, G.H. Dufour, 1845–64.

53 Hachures were common on nautical charts as well as topographic maps in the nineteenth century. The detailed hachures on this 1862 Admiralty chart of the Sunda Strait, Indonesia, had to be re-engraved after the volcanic eruption on Krakatoa in 1883.

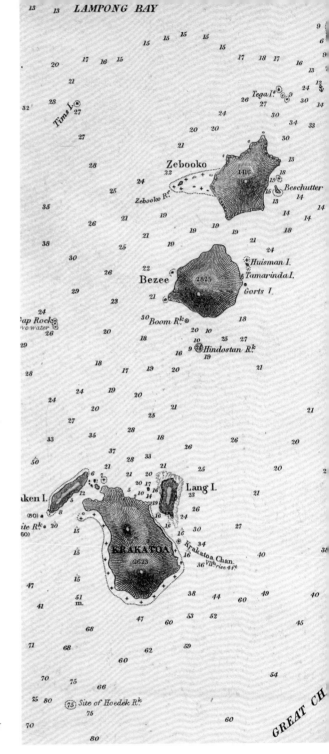

54 Hachures are the dominant form of relief representation on this map, but here they are complemented by subtle contours which give a more accurate picture of the height and form of the landscape. Extract from *Arran*, Ordnance Survey One-inch, 3rd Edition, 1911.

A greater level of precision led to a more systematic approach to hachures being established by the Austrian Johann Georg Lehmann in 1799, an approach which was subsequently adopted by several countries for their military maps and their national topographic map series. He devised *slope hachures*, which use the length, thickness and spacing of lines to represent the steepness and direction of slopes. The steeper the slope, the closer and thicker the lines. This imagines that the landscape is lit from directly above, and portrays the amount of light reflected back from the terrain. Lehmann defined nine categories of slope depending on this reflectance. Flat land would reflect all the light back vertically and so would be devoid of hachures and appear white, while slopes greater than 45 degrees would send all the light to the side and so would appear black. Varying line thicknesses define the hachures for the categories in between.

A refinement of Lehmann's technique varies the imagined angle of light from vertical to oblique, generally from the north-west (top left), and anticipates where shadows of high ground fall. *Shadow hachures* use thinner lines on the illuminated slopes and thicker lines on the shaded slopes. This subtle distinction between the illuminated and shadow sides allow the three-dimensional nature of the terrain to be perceived – an early forerunner to hill-shading techniques. Swiss cartographers, long-time masters of relief representation, perfected this technique with their national map series known as the *Dufourkarte*, produced between 1845 and 1864 (fig. 52).

Hachures aim to give an impression of the overall topography of an area. Their obvious limitation is the lack of height information, but through careful combination with spot heights and contours this can be shown. One happy accident of hachures is that lower, flatter areas have fewer hachures, so that the depiction of general map information – settlements, roads, and so on – is easier and clearer.

Despite further technological developments in mapping, particularly three-dimensional visualization techniques (see p. 199), the use of hachures persists. There is still room in cartography for the more subjective and artistic methods of relief representation which rely on personal interpretations of the landscape rather than on detailed survey data.

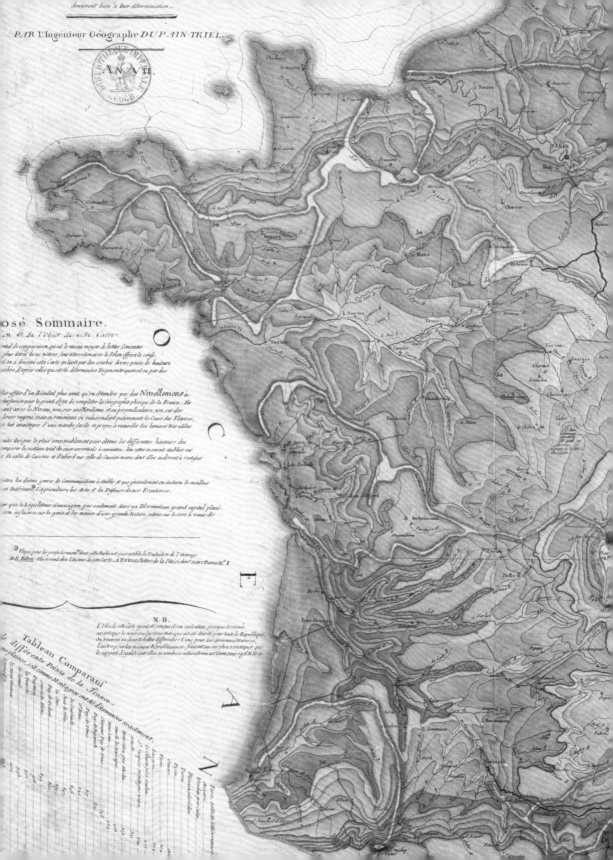

CONTOURS & ISOBATHS
On the level

Why and how would somebody want to weigh a mountain? This is what a group of scientists from the Royal Society set out to do in 1774. If they could measure the mass and density of a single mountain, they thought, they could then calculate the same things for the whole Earth and even for other planets, the Moon and the Sun. The mountain they chose – because of its roughly symmetrical shape – was Schiehallion in Perthshire, Scotland, and the man carrying out the measurements was mathematician and surveyor Charles Hutton. From a laborious survey of the mountain, a confusing mass of thousands of spot heights were measured. To make sense of them, Hutton struck upon the idea of joining up points of equal height, giving birth almost by accident to the concept of contour lines.

Hutton's lines may have been tentatively sketched, but today contours are a confident and indispensable part of topographic maps, derived from sophisticated three-dimensional measurements of high-resolution aerial photographs. Eduard Imhof, the Swiss doyen of all mountain cartographers, claims that the contour is 'the most important element in the cartographic representation of the terrain'.[5]

Contours depict lines of specified heights (usually in metres) above a defined point or datum, and have a fixed height difference between them,

55 Dupain-Triel was the first to use systematic land contours and, later, layer colours on a general map. This version of his map combines both methods. Extract from *Carte de la France ... par une nouvelle méthode de nivellement...*, Jean-Louis Dupain-Triel, 1798–9.

56 Extract from *Map of the Merwede riverepth river*, Nicolaas Cruquius, 1729–30. The first map to use submarine, or bathymetric, contours (isobaths) to indicate depth of water. The technique was quickly adopted by other cartographers to map other rivers and the open sea.

known as the 'vertical interval' – commonly 10 m, 20 m or 50 m on topographic maps, smaller on large-scale plans, greater on regional and world maps. This interval, along with the overall complexity and accuracy of contours, will depend on the scale of the map, but certain conventions for mapping contours have developed irrespective of scale. They are generally continuous lines, with every fifth (or, more rarely, tenth) line commonly treated as an 'index contour', shown by a thicker line and usually labelled with its height. Where the contour interval is too great to indicate the pattern of the terrain, additional 'intermediate' contours are added as thinner or dashed lines. Maps of mountainous terrain may need to break lines when the slopes are too steep to show each progressive contour, while

the colour of contours may vary to indicate the nature of the ground –
conventional reddish-brown for general vegetated or built-up areas, black
for rock and blue for permanent snow or glaciers.

Although Hutton was perhaps the first to create land contours in this
way, he wasn't the first to join up points of equal elevation. Early Dutch
surveyors had shown water depth in this way – Pieter Bruinsz as early as
1584, Pierre Ancelin in 1697 and Nicolaas Samuel Cruquius, who created
the first systematic submarine contours of the River Merwede, in 1730
(fig. 56). The first to be used in the open sea (specifically the English

57 A combination of contours and hill shading provide a dramatic portrayal of the radial
drainage pattern of Africa's second-highest mountain. *Tourist Map of Mt Kenya, National Park
& Environs*, Directorate of Overseas Surveys, 1974.

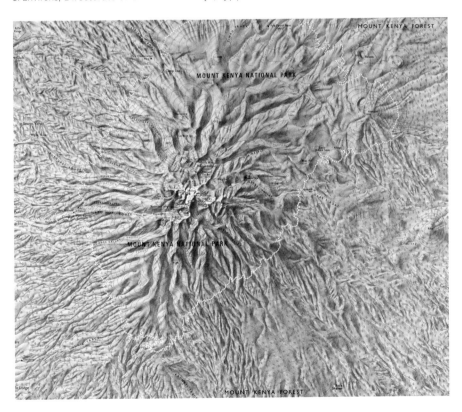

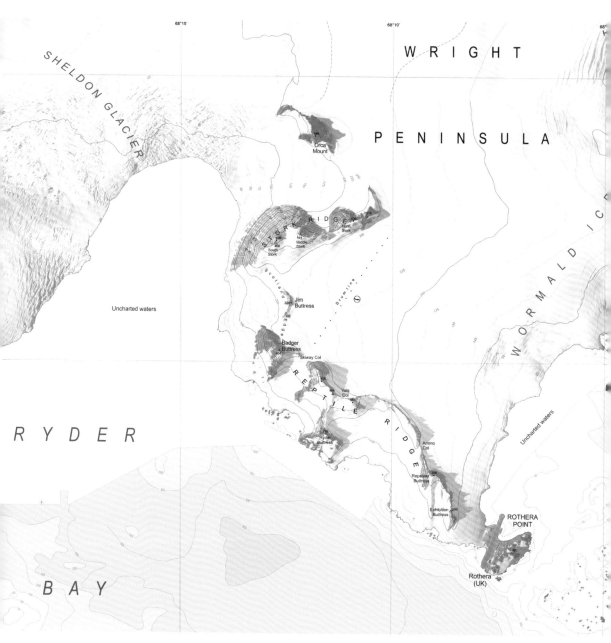

58 Based on detailed aerial photography of Antarctica, this map varies the colour of contours according to the terrain – brown for bare ground, blue for permanent snow or ice. *Ryder Bay*, British Antarctic Survey, 2007.

Channel) were drawn by Philippe
Buache in 1756. These *isobaths*
– lines measuring depth below a
datum rather than height above it
– are now fundamental features on
hydrographic charts (see p. 75).

The move to showing contours
on land lagged behind isobaths
because of the greater difficulty of
measuring land heights. Following
on from Hutton, the first general
map using contours was Jean-Louis
Dupain-Triel's 1791 map of France (fig. 55). The techniques then became
well established by the early nineteenth century. The British Ordnance
Survey approved the use of contours in 1849, although it took some time
for them to fully replace hachures.

Imhof also claims that contours are the 'basis for other types of representation'. Hachures follow contours and add terrain detail between them,
while layer colours add different shades to elevations defined by contours.
Hand-drawn hill shading uses contours as the basis for identifying which
areas appear illuminated and which fall in shadow.

They now serve a less scientific purpose than they did for Hutton and
his colleagues, but through their ability to easily indicate steep slopes
(contours close together), flatter areas (spaced well apart), valleys and
ridges (V shapes in different directions) and cliffs (contours suddenly
disappearing), contours have become the most widely understood way of
giving a map user an impression of the shape of the land.

BREAKING THE CONVENTION

Field survey, for the measurement of
contours, was laborious, particularly
in rough terrain and in tropical areas.
In 1924, in present-day Ghana, West
Africa, a team of exhausted army
surveyors had one hill left to survey. It
was a hill too far. To avoid more work
in the tropical heat, they sketched in
a fictitious contour in the shape of an
elephant. The feature remained on
topographic maps until the 1960s.

LAYER COLOURS
Rose–tinted

Visitors to the Paris World Fair in 1878 were probably surprised to find among the fine art, new technology (including Alexander Graham Bell's telephone) and machinery (a solar-powered engine was on show) a display of maps with a new look. John Bartholomew, of the famous engravers and map-makers Bartholomew & Sons in Edinburgh, was showing maps which portrayed relief in a unique way. Bartholomew's method was that of *layer colours* or *hypsometric tints*, which used colours to indicate elevation. Contours were filled in with carefully chosen colours so that a sequence of colours ran from low ground to high ground – the progression of colours making it obvious where the highest ground was.

Although Bartholomew wasn't the first to use this method, he refined it – some would say perfected it – and increased its popularity. It was used on the beautiful Bartholomew 'half-inch' maps of Britain (fig. 59), and layer colouring has been a distinguishing feature of the *Times Atlas of the World* since the Bartholomew-produced 1922 edition.

59 Extract from *Firth of Clyde,* John George Bartholomew's *Half Inch to the Mile Maps of Scotland,* 1926–35. Layer colours in the classic Bartholomew style make this series of maps one of the most attractive representations of Scotland's landforms.

60 OVERLEAF John George Bartholomew produced this collection of maps of Scotland's inland lochs. They display an attractive balance between the layer colours on land and water. *Loch Maree, Bathymetrical Survey of the Fresh-Water Lochs of Scotland,* Sir John Murray, 1897–1909.

PLATE I

BATHYMETRICAL SURVEY O

SIR JOHN MURRAY, K.C.

LOCH MAREE
(UPPER SECTION)
AND LOCH GARBHAIG
(EWE BASIN)
SURVEYED IN 1902 BY
T. N. JOHNSTON, M.B., C.M., JAMES PARSONS, B.Sc.
T. R. H. GARRETT, B.A., JOHN HEWITT, B.A.
JAMES MURRAY AND SIR JOHN MURRAY
Height of Surface of Water above Sea Level—Loch Maree, 30.3 feet
The Land Contours are from the Ordnance Survey

Scale 1 : 21120 3 INCHES TO 1 MILE   Mile 1

0·21 KILOMETRE = 1 CENTIMETRE   Kilometres 2·61

LOCH G.

LONGITUDINAL SECTION ALONG

THE GEOGRAPHICAL JOURNAL 1904

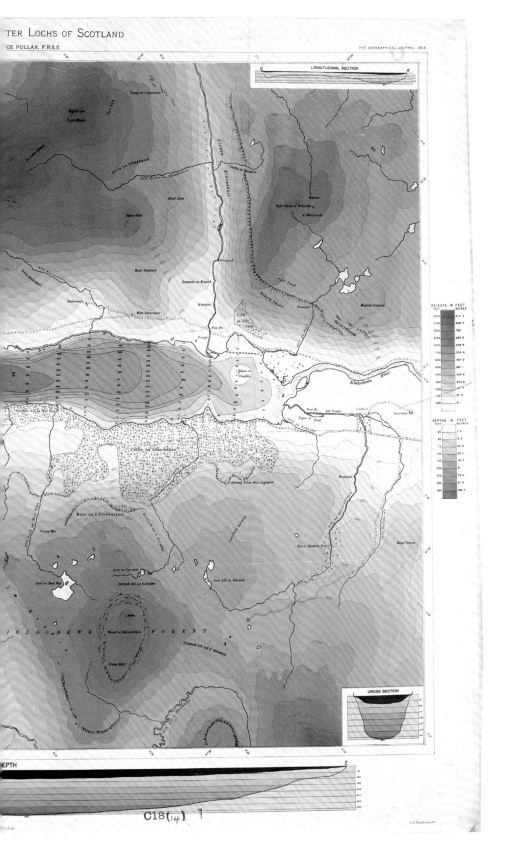

LONGITUDINAL SECTION

CROSS SECTION

HEIGHTS IN FEET

DEPTHS IN FEET

Kinlochewe River

Coille na' Glas-leitire

INLOCHEWE FOREST

Frenchman Jean-Louis Dupain-Triel first used layer colours for his 1798 map of France – an enhanced version of his 1791 map which was notable as the first general map to use contours. He used several shades of grey to indicate different elevations. The technique was also used on a map of Scandinavia by Carl Forsell (1830) and on wall maps by Emil von Sydow (1837). Colour tints based on isobaths have been used to show sea depths since around 1830. As contours became more widely adopted for military and national mapping in the early nineteenth century, their enhancement with layer colours became increasingly common.

An almost continual debate centres on the sequence of colours to use for the layers. Various conventional schemes have become established, although there is an infinite range of possible colour combinations. The representation of depths in the sea has become fairly standardized as different shades of blue – white or very light blue for the shallowest water, and progressively darker blues as depth increases. Although some maps use a similar graded approach for land areas with greens or browns, depictions of the progression from lowlands to mountains generally use a wider range of colours. Arguments raged in the early days as to whether the colour sequence from low to high ground should be light colours to dark, or dark to light. The fact that dark colours in low-lying areas, where most other map detail appears, could lead to problems of legibility contributed to the wide acceptance of light-to-dark schemes. In the late nineteenth century, sequences following the spectrum began to appear and have become the most widely accepted. Although there have been many variations on this approach (including the use of red to make the highest ground really stand out), the sequence of grey-green through greens, yellows, red-browns, to violet and white for the very highest mountains has become the norm.

In an attempt at universal map standardization (not just for layer colours but for all aspects of map design), the 1962 Technical Conference on the International Map of the World (IMW – see p. 189) adopted a version of this colour scheme. But it was soon found that one size doesn't fit all.

61 This chart presents complex bathymetric data in a beautiful light-to-dark progression of blues, with a very clear distinction between land and sea. Extract from Sheet 5.12 Fifth Edition of the *General Bathymetric Chart of the Oceans (GEBCO)* by the Canadian Hydrographic Service, 1995.

Different environments and types of terrain demand different approaches, and there is a danger that having low land always shown as green can imply, obviously incorrectly in some parts of the world, that those areas are richly vegetated. Layer colours are not an indication of land cover, but only of elevation. Complications can also arise when other methods of relief representation are used in combination. Hill shading will affect the consistency of layer colours, as will the extensive use of hachures.

Although the response to the Bartholomew maps at the Paris Fair was mixed (some thought the techniques dumbed-down cartography), they were awarded an 'Honourable Mention'. They have certainly been much admired since, and the approach they took has helped establish layer colours as the most effective way of giving an at-a-glance impression of the relief of an area.

HILL SHADING
Out of the shadows

What do watches, chocolate and Roger Federer's single-handed backhand have in common? They are all expressions of the widely recognized Swiss characteristics of precision and perfection. Swiss cartographers have applied these traits to their maps, and topographic maps of Switzerland are widely accepted as the most beautiful. In particular, since the early efforts of Hans Conrad Gyger in 1664, they have mastered the art of hill shading.

This method of relief representation relies on a perception of the three-dimensional landscape through the use of light and shade. It is based on the idea of illuminated landscape models – real or, more generally, imagined. From the late nineteenth century physical models were constructed from plaster, carefully illuminated and photographed. The shadowed image was then combined with a base map of the area. Today, the 'model' is imaginary or digital. For hand-drawn hill shading, an interpretation of the effects of how light falls on the terrain is rendered artistically in pencil, watercolour or airbrush ink. Digital hill shading uses lighting effects on detailed digital terrain models consisting of dense networks of precisely measured elevations. While being arguably more 'accurate' because of the data it uses, digital hill shading does not yet quite match the beauty of

62 *Appenzell (School Map)*, Eduard Imhof, 1923. This map displays Imhof's mastery of reading and representing how light can bring out the relief of a mountainous region. His hand-drawn shading, which also incorporates subtle layer colours, brings the terrain to life.

63 The earliest origins of cartographic hill shading are seen in *A Bird's-Eye Map of Western Tuscany*, Leonardo da Vinci, *c*.1503–4. Although not in plan view, the characteristics of the hills depend on careful representation of how light falls on them.

hand-rendered shading – perhaps precisely because it uses mathematical processes rather than artistic expressions.

The degree of contrast between light and dark, highlight and shadow, is critical to its success, and this principle – known in art as *chiaroscuro* – was used on a map of Tuscany, Italy, by Leonardo da Vinci as long ago as 1503 (fig. 63). Da Vinci's map showed hills in a perspective or a 'bird's eye' view (see p. 199), rather than plan view, but the technique was subsequently applied to maps, notably by William Roy and Paul Sandby on some of the first military maps of Britain in 1791. These assumed the map to be illuminated from directly above, with the amount of reflected light determining the shading applied – the steeper the slope, the darker its shading.

Two developments in particular revolutionized the art of hill shading. The printing process of lithography, invented in 1798 and refined by the use of 'halftone screens' in 1852, allowed continuous tone images (for example, photographs and original artwork) to be printed and the subtlety of tonal variations to be reliably reproduced. In the same period, the practice of illuminating the imagined landscape from an oblique angle rather than from above became established. This allowed a clearer impression of topography to be produced, emphasizing the distinctions between those features that faced the light, and those that lay in shadow. Infinite tonal variations could then be used to portray slopes in relation to how the light fell on them.

The direction of light used to create the effect is critical. Although conventions for hill shading are rare, one is that the light comes from the north-west (or top-left). It seems odd, though, that this is opposite to the general direction of sunlight, in the northern hemisphere. Although people's perception of hill shading can vary, it is common for the effect of the shading to be reversed if the light comes from the south-east, with mountains appearing as valleys and vice versa. A single light direction can mean that features aligned with the light are not evident in the shading. The Swiss style of hill shading, in particular, uses slight local adjustments of the light direction to ensure that all characteristics of the terrain are brought out.

Hill shading usually only depicts the relief and doesn't take account of the nature of the land cover. Tones of grey are the most common colours used, although some of the best examples use pale violet in the shaded areas and yellow or white to add highlights to the illuminated side. For specific landscapes, colours relating to the terrain may be used – sandy browns and buffs for desert areas, greens and browns for the progression from farmland to upland – but too many variations can lead to misinterpretation and get in the way of the main task of showing relief.

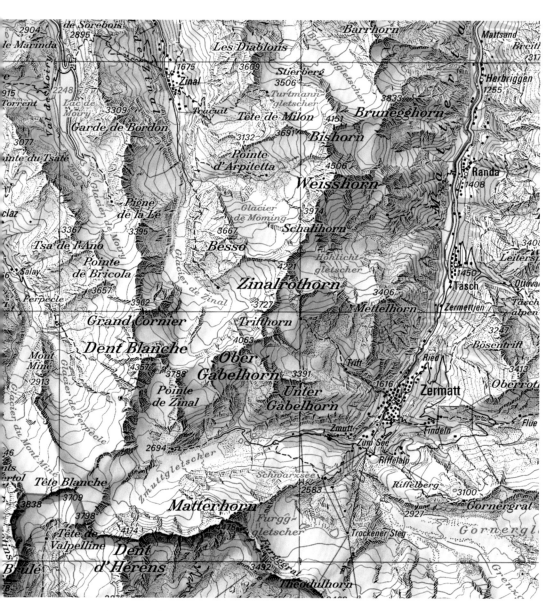

64 *Extract from Swiss National Topographic Map*, Federal Office of Topography swisstopo, 2018. Modern Swiss topographic maps continue the traditions of dramatic hill shading. Detailed rock drawing and coloured contours, reflecting the nature of the terrain, complete the effect.

65 Hill shading may be generated by photographing illuminated physical three-dimensional models. The effect is similar to that seen on this large relief model of Iceland to be found in Reykjavik City Hall.

When combined with contours and with rock drawing – another technique perfected by Swiss cartographers – hill shading is an attractive and
effective means of depicting a landscape, with the best examples being the
perfect blend of art and science in map-making.

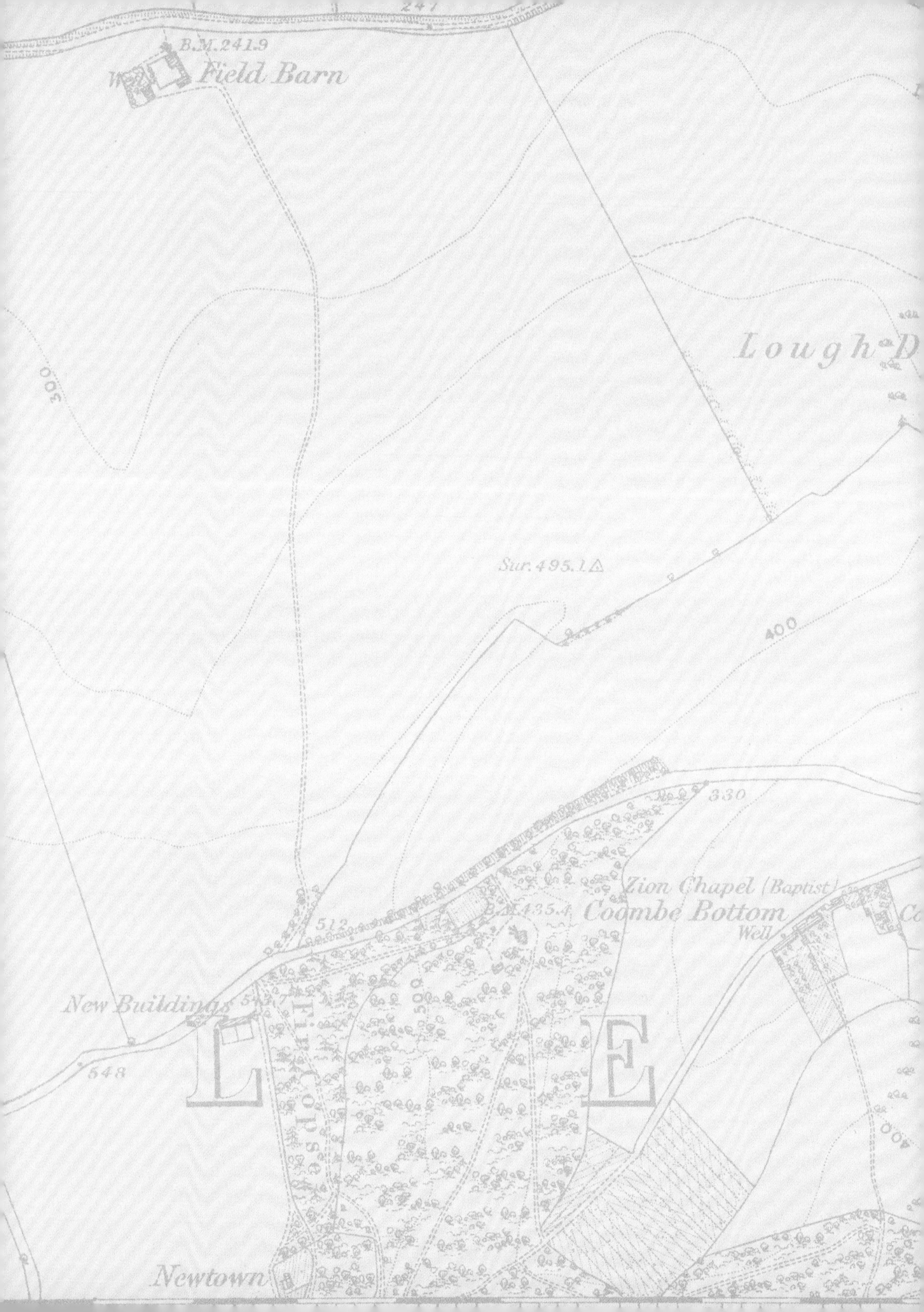

B.M. 241.9
Field Barn

Lough°D

Sur. 495.1 △

400

330

Zion Chapel (Baptist)
Coombe Bottom
Well

512

B.M.435.4

New Buildings

548

Newtown

Part IV
NAMES & BOUNDARIES

PLACE NAMES
Putting a name to it

PARIS *France* city 48° 52′ N, 2° 21′ E; 56 B3

A typical atlas index entry such as this aims to answer the question, 'Where is it?' Most people using an atlas or map will be looking for a specific place or geographical feature and it is this element of 'place', along with how such places are named, which presents map-makers with unique challenges.

Maps have shown place names since ancient times. Town plans on clay tablets from Mesopotamia, such as that of the town of Nippur in *c.* 1500 BCE, used cuneiform characters to identify features (fig. 67), and maps from the Chinese Han dynasty (206 BCE–220 CE) named settlements and rivers. Place names are an integral part of general or topographic maps. But they are a sensitive subject, at both local and international levels. Invariably, many letters of complaint received by map publishers, or social media comments on the inaccuracy of online maps, will relate to place names.

Whichever names are shown on a map – and this depends on the purpose of the map and judgements about which places and features are important enough to be shown – the relationship between each name and its corresponding feature needs to be unambiguous. While the position of place names is by definition largely defined by geography, many

66 This extract from a map of Oxford from 1973 was part of a worldwide Soviet mapping programme carried out during the Cold War. The Cyrillic place names would need to be transliterated into the Roman alphabet for locals to make sense of it.

conventions have developed around the issue of name placement. In general, names will be aligned horizontally to the frame of a map, or, in the case of small-scale maps, will align with lines of latitude (parallels). This varies for linear features (rivers, mountain ranges, and so on), for which names will follow the alignment of the feature. Preferred positions for settlement names relative to their symbols have emerged, and are now built into software used in the compilation of maps. In Western cultures, the preferred position for a settlement name is to the right, and slightly above the level of its symbol. The density of place names in a particular area will not always allow this, and, as with many cartographic conventions, compromises need to be made. The

67 Babylonian clay tablet showing a map of the ancient city Nippur, *c.* 1500 BCE. Names of places and features appear in cuneiform script. Such maps may have been for defence or administrative purposes, or to record land ownership.

aim for names of areal features such as countries, lakes and forests is to keep the name entirely within the feature – the more a name crosses other map detail and linework, the less legible it becomes.

Even if a name is placed perfectly, is it 'correct'? Although vast databases of geographic information are now at a cartographer's disposal, the spelling of place names is not always clear-cut. One of the excitements of international travel is discovering that places have their own local name form – you may leave on a flight bound for Vienna (its 'exonym' or conventional English name) and land in Wien (its local German name form). A map will commonly use either local name forms or their 'conventional' equivalents in the language of the publication. This is complicated further by the fact that not all languages use the Roman alphabet. Place names in countries using Cyrillic, Chinese or Arabic scripts, for example, need to be converted into conventional forms in the Roman alphabet by transliteration or transcription (fig. 66).

Incorrect treatment of names can cause serious geopolitical issues. Naming a feature in a particular way can imply ownership, and as soon as somebody else comes up with a different name form hackles may rise. Great sensitivity surrounds names such as the Persian Gulf (which Arabic countries call the Arabian Gulf) and the Sea of Japan (the preferred name form in Japan, to the great consternation of Koreans, who call it the East Sea and regularly lobby others to do the same).

Names can also change over time. Names dating back to difficult times, or referring to notorious past leaders, may go out of favour, ancient names used by indigenous peoples may become fashionable, and changes in linguistic preferences may also lead to 'new' names being adopted – Kolkata is now the widely accepted name form for the Indian city formerly know as Calcutta. Depending on the purpose and scale of a map, a cartographer can cover several options by following the convention of showing alternative names in brackets.

In the UK and the USA, official organizations aim to address such issues. The Permanent Committee on Geographical Names for British Official Use (PCGN) advises the British government on preferred name forms throughout the world, with the US Board on Geographic Names (BGN) doing similarly for the authorities in the USA. Both organizations issue strict guidance on how to transliterate non-Roman place names. The United Nations Group of Experts of Geographical Names (UNGEGN) also plays a part in working towards international standardization of name policies and resolution of disagreements between nations.

The seemingly straightforward question 'Where is it?' involves more than just geographical location. There may be various interpretations of what 'it' is known as, and place name conventions are not universal. Several map publishers have found out to their cost that the choice of place names is not always straightforward, and there are those who will quickly make it clear to them and to others when a 'wrong' choice has been made.

BREAKING THE CONVENTION

Names of places can change for various reasons, but when a popular US radio show was looking for a town to change its name to the name of the show, Hot Springs, New Mexico, jumped at the chance. For the privilege of hosting the show's annual 'fiesta', and a chance to distinguish it from other settlements called Hot Springs, it was renamed 'Truth or Consequences' in March 1950.

68 This extract from a portolan chart of the Adriatic, c. 1400, is from a time when most known places were ports. Place names positioned perpendicular to the coastline give these maps a distinctive look and almost define the shape of the land themselves.

BOUNDARIES
Drawing the line

It looks so neat. Each country precisely drawn, with clear lines and an attractive line of colour between them, the world map can look both pleasing and peaceful. But look more closely, perhaps by zooming in on Google Maps, and it will soon become clear that the boundaries between countries aren't as clear-cut as they seem. In some parts of the world the pattern of international boundaries is actually fiendishly complex. So complex, and sensitive, in fact, that a single representation on a map may not be enough. A user of Google Maps in India, for example, would see a different picture of that country's boundaries than a user elsewhere. Similarly, log on in China and you will see a pattern of boundaries unique to the Chinese view of the world.

By their nature, boundaries are geographical features and so maps have an important role to play not just in showing where they are, but also in defining them in the first place. And this doesn't involve just international boundaries. Administrative divisions, local parishes and individual properties all need to be mapped to allow governments to function, and disputes to be resolved or avoided.

Boundaries have been a challenge for map-makers since ancient times. Even in ancient Egypt, Mesopotamia and China maps showed property

69 *A Map of the British and French Dominions in North America...*, John Mitchell, 1755. What is known as the 'Red Lined' map was used to negotiate and define the territorial boundaries established by the Treaty of Paris of 1783.

POLITICAL MAP
OF
AFRICA

BY J.G. BARTHOLOMEW F.R.G.S.

REFERENCE TO COLOURING

BRITISH   PORTUGUESE   BELGIAN
FRENCH   ITALIAN   TURKISH
GERMAN   SPANISH   AFRICAN STATES

FORESTS AND RAILROADS
SALT LAKES AND PANS

THE BRITISH ISLES
on same scale as Map of Africa

and territory boundaries. An estate map of Nuzi in Mesopotamia, drawn on a clay tablet, dates from around 2500 BCE. Cadastral maps – plans for taxation purposes – go back to Roman times, when maps also depicted the frontiers of the Roman Empire. The earliest boundary map from England was of Sherwood Forest in the mid-twelfth century, but it was in the fifteenth century, when the idea of the nation-state was evolving, that mapping boundaries became more important, more commonplace and increasingly complex.

One of the earliest maps representing a contemporary geopolitical issue is the *Cantino Planisphere* of 1502 (see fig. 6). This shows the situation arising from the signing of the Treaty of Tordesillas of 1494 which basically divided up the New World between Spain and Portugal. Maps also played a crucial role in the establishment of the United States of America. John Mitchell's 1755 map of Britain's American colonies (later referred to as the 'Red Lined' map) was used during the negotiation of the 1783 Treaty of Paris, and is perhaps the most important map in the history of North America (fig. 69).

Mention of treaties makes things sound straightforward, but the definition and mapping of boundaries can be anything but. Boundaries need to be agreed (though this may be difficult if a boundary has been imposed from a distance by some superior authority) and then described in general terms – overall location, geographical coordinates (taken from maps) and alignment relative to known points on the ground. They then need to be delimited on maps through detailed interpretations of the treaty or agreement, and finally demarcated on the ground, commonly by the installation of boundary markers, which again need to be accurately mapped. However, things are rarely this clear-cut and this theoretical process may be broken at any stage. Numerous types of boundary exist as a result of the success or failure of this process. Boundaries may be agreed and established in law

70 This *Political Map of Africa*, J.G. Bartholomew, 1898, reflects the 'Scramble for Africa' which was triggered by the Berlin Conference in 1884. The Conference resulted in the colonial powers establishing a new pattern of boundaries across the continent.

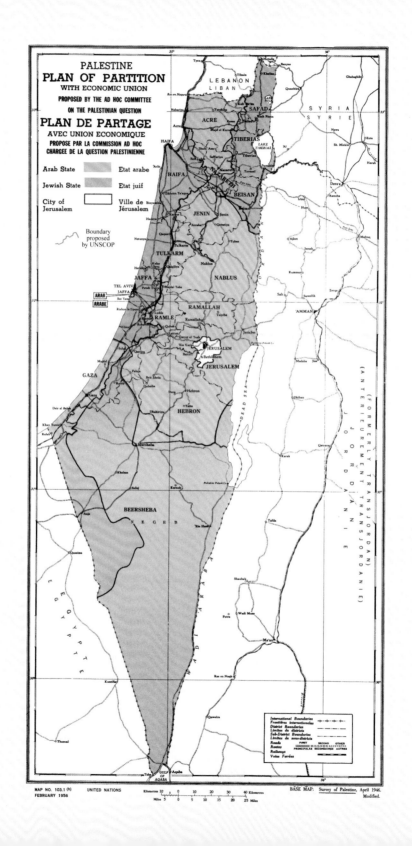

– *de jure* boundaries; their precise alignments may be in dispute; they may be agreed but not demarcated; they may be temporary ceasefire lines; or they may be so well established that they have become settled in the eyes of the international community without going through the normal procedures.

So how do cartographers deal with these complexities and show boundaries in an authoritative and clear way? By definition, boundaries are linear features and have always been represented on maps as line symbols (see p. 69). A simple line may not allow the full story of a boundary to be shown, but certain conventions go some way towards explaining things. Solid lines are often used to show uncontested and established boundaries, while any combination of dashes and dots can depict disagreements or inconsistencies in disputed boundaries. Subtle colour bands along boundaries – which were used by Christopher Saxton on his county maps of England as early as the sixteenth century and on the earliest maps of the United States – have also become an established method of highlighting geopolitical divisions.

Whichever conventions are followed, the complexities of geography and geopolitics cannot be avoided. Both the establishment and the subsequent representation of boundaries on maps are liable to be highly sensitive and have far-reaching consequences. Maps have featured prominently in many conflict situations around the world – the Oslo Accords in the Middle East; during the Partition of India; in the definition of the Dayton Line in the Balkans (see fig. 18); throughout the decolonization of Africa – either as a cause of conflict (perhaps through the arbitrary definition of a boundary by a distant colonial power) or as a means of resolving a crisis. The representation of boundaries remains a sensitive issue; if a wrong choice is made in where to draw the line, sensitivities may be aroused and the map may become a catalyst for further dispute.

71 *Palestine Plan of Partition*, United Nations, 1946/1956. Maps such as this played a part in the negotiations around the partition of Palestine and the establishment of the State of Israel in 1948. The apparent simplicity of boundaries disguises many political complexities.

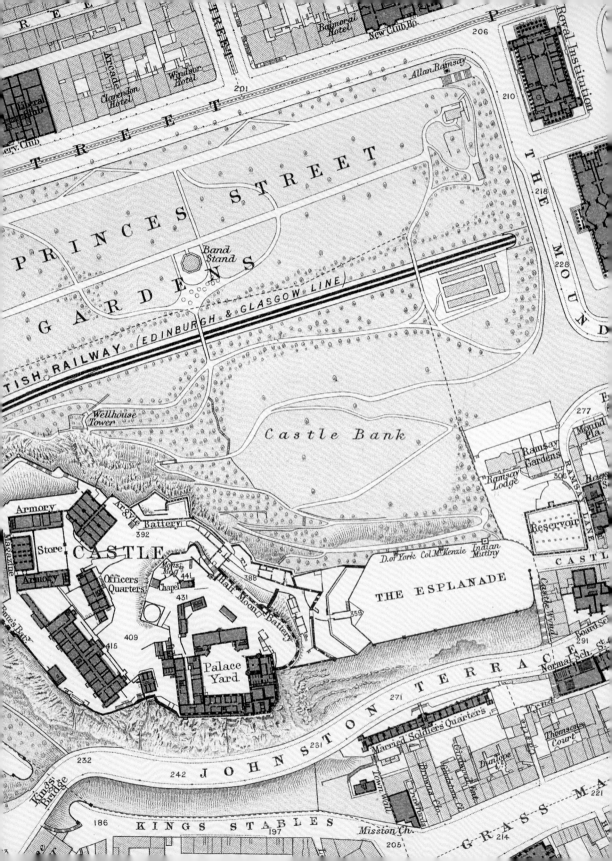

TYPOGRAPHY
Keep it clear

When one of the greatest cartographers of all time published a typographic manual in 1540, it may have seemed like a strange departure from his day job. But in producing his *Literarum latinarum*, Gerard Mercator could have been making a statement about the importance of lettering in map design. The work is a detailed guide on how to write italic, or more specifically chancery cursive script – a style of writing first used on maps by Benedetto Bordone in 1528 – and was aimed at ensuring legibility. As a standard guide to modern cartographic techniques states, 'Perhaps no element of a map is so important in conveying an impression of quality as the styles and sizes of the type employed and the manner in which the map has been lettered.'[6] Methods for producing type on maps have changed, but recognition of the need for clarity certainly goes back to Mercator's time.

Words, in particular place names (see p. 131), have appeared on maps since the original development of written languages, and the styles of lettering on maps through the ages reflect developments in typography – the art of creating and arranging type – and the technologies involved in applying type to maps.

Until the fifteenth century, text for virtually any map (as for any type of publication) was produced by hand – in manuscript. Each map was

72 *Plan of the City of Edinburgh with Leith & Suburbs*, Bartholomew & Son, 1891. Bartholomew were masters of copper engravings of maps, and this extract highlights the beauty of engraved type, with varying styles and forms, and perfect letter spacing.

73 An early engraved map showing greater precision and consistency, using a good variety of type styles and forms, including serif, upright and italic, and capitals and lower case to distinguish between features. Extract from *Map of Lothian, Scotland*, J. Blaeu, c.1654.

uniquely drawn, and the creation
of multiple copies was a laborious
and error-prone task. Various styles
of manuscript lettering developed
(including cursive script), and were
commonly applied to maps, but
consistency of lettering was very
difficult to achieve.

In China, in 1155, a map was
created as a woodcut print. This
method of reproduction involved
the carving of a block of wood to
leave a raised image area, which
was then inked and transferred
to paper (a technique familiar to
anybody who has tried lino printing). Although initially creating
fairly coarse lettering, it ensured
some degree of consistency. The
earliest woodcut map to appear in
a printed book was a simple world map (known as a 'T-O map' because of
the basic shape it uses to represent the Earth) originally created by Isidore
of Seville in the seventh century as part of his *Etymologiae*. This work was
later printed as a woodcut in Augsburg, Germany, in 1472 (fig. 74).

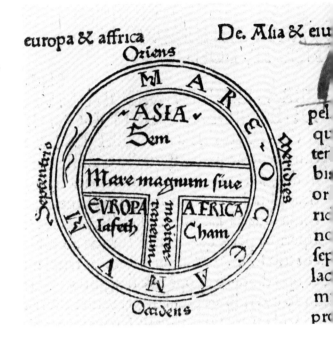

74 This map is in the characteristic shape of
medieval 'T-O' world maps, or *mappae mundi*,
and has more type than map detail. Copy of
a world map from Isidore of Seville's seventh-
century work *Etymologiae*, 1472.

The desire for greater precision and variety of lettering styles was met
by the development of engraving in the fifteenth century. In contrast to
woodcuts, the image area on an engraving consists of incisions made by
special tools into a metal plate (usually copper). The incisions, which are
often incredibly fine, then carry the ink for the transfer of the image to
paper. Engraving developed throughout the next 300 years and remained in
use for maps well into the twentieth century.

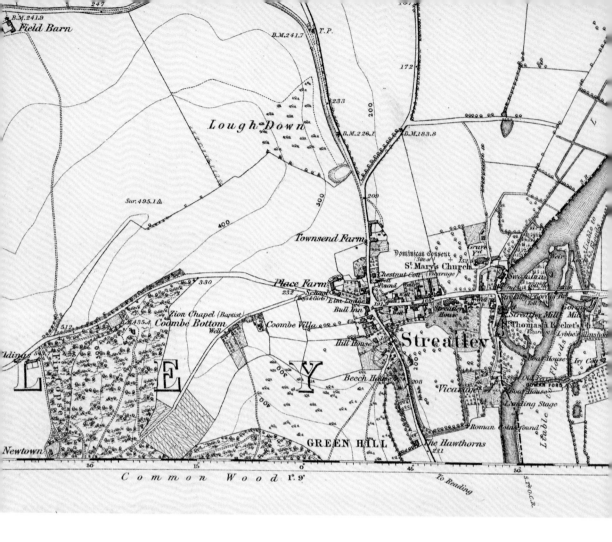

75 The unusual treatment of the perfectly engraved upper-case letters on this map ensures that they don't dominate the map but lie 'beneath' the topographic detail.

Technological developments, including movable type (individual letters cast in metal which are manually positioned to create and arrange words), phototypesetting, and most recently digital type led to huge changes in the appearance of text on maps, and took us ultimately to today's abundance of fonts for cartographers to choose from.

Maps have always presented unique typographic challenges compared to books. Place and feature names need to be combined with graphic elements of the map (whose positions can't change), need to be accurately positioned and often need to be small, yet remain legible. The choice of font and the positioning of text are crucial to the overall appearance and functionality of a map. Certain conventions on the use of type have developed to enable these things to happen, but these are not universally applied and the treatment of type can go a long way towards judging how good a map actually is.

Typographers speak in terms of a font's *style* and *form*, and cartographers are well advised to consider these characteristics carefully in deciding how to annotate a map. Careful choices will not only ensure legibility but also allow the user to distinguish types of features and their relative importance. A font's style generally defines it as serif or sans serif (with or without the small extensions at the end of individual strokes of letters); 'old style', 'modern' or 'decorative'. Form describes type in terms of it being roman (upright) or italic; normal width, condensed (with narrower characters) or extended (widened characters); capitals or lower case and whether the weight of individual letters is bold, medium or light. These variables give the map-maker some powerful design tools.

Conventions have developed which often govern the use of fonts to distinguish between different types of feature on a map. Most commonly, physical features are named with italic type and man-made or cultural features with roman type. Further visual distinction is possible with the use of colour – blue type for water features, for example. The relative hierarchy of features (for example, town symbols representing settlements of different populations) is commonly portrayed by varying the size and weight of type, and the use of capital letters.

Mercator would no doubt marvel at the range of typefaces and the technology now available, but he would also perhaps recognize that the challenge of choosing type styles whilst retaining legibility among complex geographical detail remains.

NORTH

CALIFORNIA

MARYLAND
VIRGINIA
CAROLINA
JamesTown
CharlesTown
C Hatteras
FLORIDA
GOLFO DE
MEXICO
NEW
SPAIN
BAY OF
CAMPECHE
CUBA
BAHAMA
ISLES
Havana
HISPANIOLA
S Domingo
P. Rico
MEXICO
Vern Cruz
YUCATAN
HONDURAS
Gulf Honduras
NICARAGUA
JAMAICA
CARIBBE
ISLES
Barbados
Panama
Darien
S MARTHA
VENEZUELA
GUIAN
North Cape
R Amazones

MAR DEL
ZUR
SOUTH
C S Francisco
C S Helen
Guajaquil
Payta
PERU
Lima
S Miguel
Arequipa
WILDE
BRAZILE
Parnibe
Pernambuc

MARE

A Correct CHART of the
TERRAQUEOUS GLOBE
According to Mercator's, or more
properly WRIGHTS Projection:
On which are descri'd Lines,
shewing the Variation of the
Magnetic Needle according
to observations made
about the Year 1756.

Sold by W. Mount
and T. Page on Great
Tower Hill London.

PACI

to shew the Inclination of the

FI
AME
CHILI
PARAGUAY
Terra dos
Patos
RI
CA

Rio de Janero

CUM

Baldivia

COSTA
DESERTA

PATA
GO
NIA

C Blanco
Pepys Isle

East Variation

Falkland I.

TERRA DEL
FUEGO

Magellan Straight

WESTE

OCE

West Variation

OCEAN

Bermudas

East Variation

West Variation

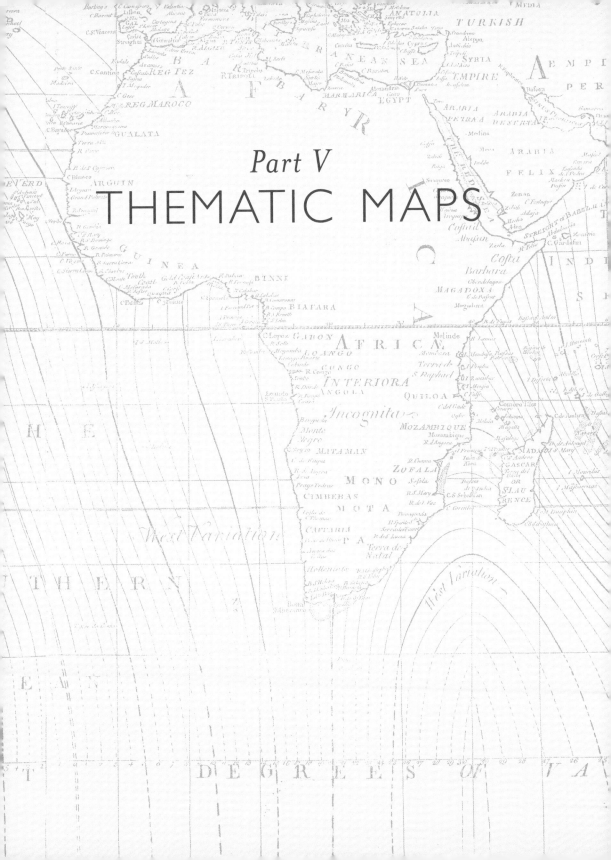

Part V
THEMATIC MAPS

QUALITATIVE THEMATIC MAPS
What do we have here?

At the time of a major outbreak of cholera in Soho, London, in 1854, in which around 500 people died in just five days, the general belief was that diseases were transmitted through bad air and unpleasant smells – a miasma. A doctor called John Snow questioned this miasmatic theory. He felt that something had to be ingested for a disease to spread and suspected contaminated water. At the risk of his own health, he plotted on a map of Soho where deaths had recently occurred (fig. 76). The incidence of diseases had been mapped since the late eighteenth century, but by adding a single feature – the location of the water pump on Broad Street, on which the local residents depended – his map became revolutionary. By visually comparing the locations of deaths with the position of the pump it became apparent that more deaths occurred in the pump's immediate vicinity. Snow was convinced the water from the pump must be the cause and persuaded the local authority to remove the pump's handle. The number of cholera deaths in the area subsequently decreased dramatically.

Maps are commonly divided into two basic types: *general* and *thematic*. General maps, which include topographic maps, show many different phenomena, without emphasizing any particular information. Thematic

76 Extract from *Map of Soho, London* from *On the Mode of Communication of Cholera*, John Snow, 1855. A groundbreaking thematic map, focusing firmly on plotting and analysing the precise locations of cases of cholera, with only minimal 'background' information.

THE STREETS ARE COLOURED ACCORDING TO THE GENERAL CONDITION OF THE INHABITANTS, AS UNDER:—

| | Lowest class. Vicious, semi-criminal. | | Very poor, casual. Chronic want. | | Poor. 18s. to 21s. a week for a moderate family. | | Mixed. Some comfortable, others poor. | | Fairly comfortable. Good ordinary earnings. | | Middle class. Well-to-do. | | Upper-middle and classes. We |

A combination of colours—as dark blue and black, or pink and red—indicates that the street contains a fair proportion of each of the classes represented by the respective colours.

77 *Map Descriptive of London Poverty*, William Booth, 1898–9. An early example of a thematic 'overlay' on a topographic map. Different colours denote 'the general condition of the inhabitants', from wealthy 'Upper-middle and Upper classes' to the 'Lowest class. Vicious, semi-criminal', according to the key.

maps (a term first used in 1953), as their name suggests, focus on portraying a specific theme or subject. This theme takes clear priority on the map, with more general information – roads, settlements, and so on – sitting in the background.

Snow's map is an early, and hugely influential, example of a thematic map which shows simple *qualitative* information. What is mapped is not given any statistical value or hierarchical significance, but is merely represented by simple symbols in their correct locations. This approach works with any phenomenon, and there are an infinite number of subjects which could be mapped in this way – everything happens or occurs somewhere and so is mappable. There is no clear point at which a general map becomes a thematic map, but as soon as a subject is identified and given clear visual priority then the map is thematic. Google Maps can provide you with a general map of your local area, but as soon as you search for coffee shops, for example, the bright drop-pins that appear make the map thematic.

Some of the earliest thematic maps were those of ocean currents by Athanasius Kircher, in 1687, and astronomer Edmond Halley's maps showing winds (1686) and variations in the Earth's magnetic field (1701). Maps even earlier than this sometimes included elements of specific themes, but can't be described as truly thematic. A map of Lancashire from 1590, for example, while general in its nature, marks with a cross the residences of Catholic families who it was feared may side with Spain in the event of a Spanish invasion. Geological maps – qualitative thematic maps showing the occurrence of rock types (see p. 167) – initially developed through the eighteenth century, but it was in the late eighteenth and the nineteenth centuries that thematic mapping flourished. This was a time when more and more mappable scientific and social data was becoming available – national censuses were established, the natural sciences were revealing new phenomena, knowledge of diseases was developing.

So how to show all this information? Cartographers think in terms of three basic types of map symbol (see p. 57) – points, lines and areas. All phenomena, natural or man-made, can generally be mapped by one or other of these types and qualitative thematic maps make good use of them all. Add to these the options of varying a symbol's size, shape and colour, and the map-maker has quite a range of tools to portray the chosen theme in a clear and unambiguous way. Throughout the design process, the theme must not be lost, and it is the measure of a thematic map whether its story is clear. Overcomplicated pictorial symbols, inappropriate colours and attempts to include too many phenomena can quickly make a thematic map difficult to understand.

Simplicity is key to the success of a thematic map. Not many will be as revolutionary, or as simple, as Snow's cholera map, but focusing on specific phenomena and mapping them carefully are a powerful way of revealing things which would otherwise remain hidden.

78 This map uses areas of colour to identify different categories of land tenure (land sold or under lease, etc.) but takes no account of the local indigenous population. Extract from *Map of the Middle Island of New Zealand*, Waterlow & Sons, 1878.

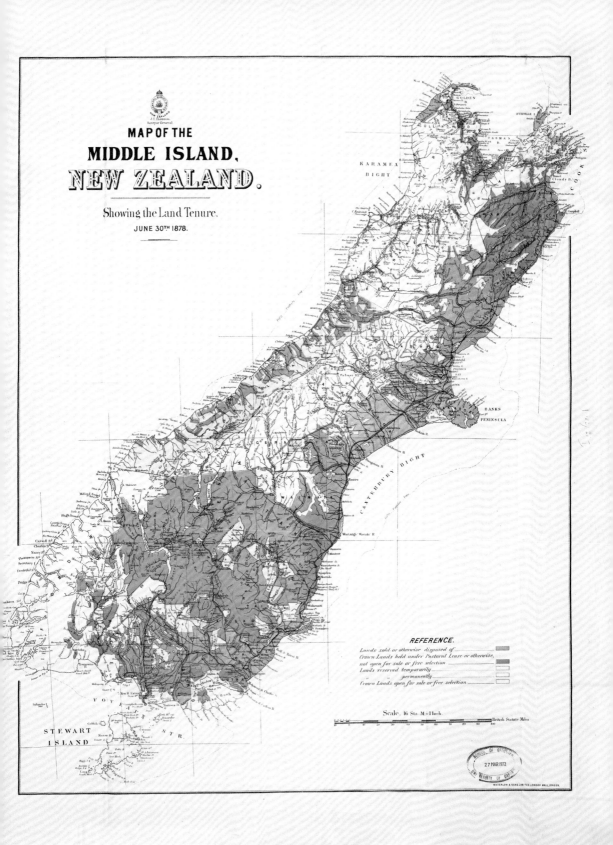

MAP OF THE
MIDDLE ISLAND,
NEW ZEALAND.

Showing the Land Tenure.
JUNE 30TH 1878.

REFERENCE.

Lands sold or otherwise disposed of
Crown Lands held under Pastoral Lease or otherwise,
not open for sale or free selection
Lands reserved temporarily
permanently
Crown Lands open for sale or free selection

Scale. 16 Sta. M.=1 Inch.
British Statute Miles

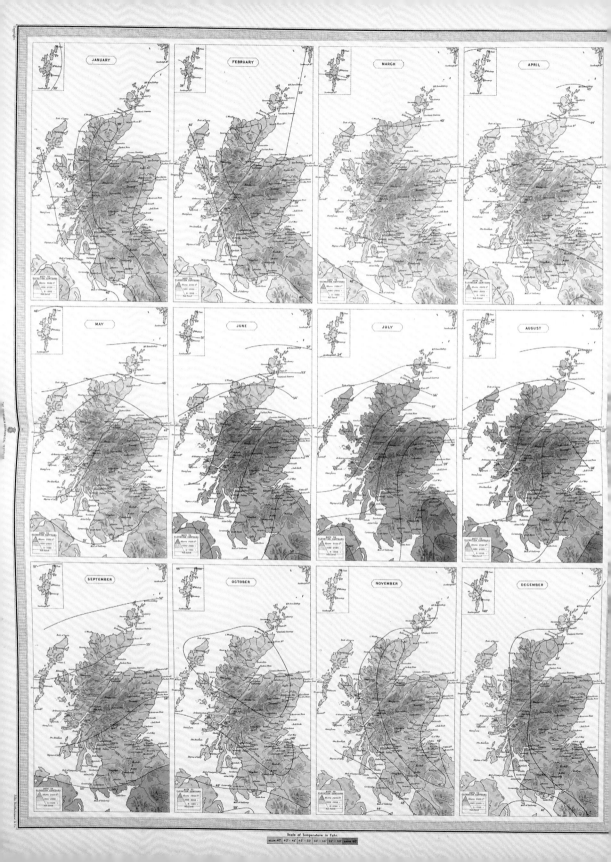

JANUARY   FEBRUARY   MARCH   APRIL

MAY   JUNE   JULY   AUGUST

SEPTEMBER   OCTOBER   NOVEMBER   DECEMBER

Scale of Temperature in Fahr.

QUANTITATIVE THEMATIC MAPS

How many are there?

On 27 March 2011 all households in the United Kingdom completed a Household Questionnaire as part of the national census. Details of the country's population – noting such details as age, gender and marital and employment status – are required by the government to plan and budget for the provision of public services, and have been collected every ten years since 1801. Some countries began similar counts even earlier – Sweden was the first to implement a systematic national census in 1749 and the United States began the process in 1790.

These censuses have generated huge amounts of socio-economic data, all of which has a geographical, and therefore mappable, element. They were established at a time of rapid development in global exploration, science and social studies. Scientific study was moving from the general area of 'natural philosophy' into numerous specialisms – meteorology, geology, biology and the social sciences, for example – and all of these were producing large amounts of information which needed to be processed and presented in some way. Although statistical techniques had become established during the eighteenth century, methods of portraying data

79 Monthly temperature maps of Scotland, from *Survey Atlas of Scotland*, J.G. Bartholomew, 1912. The growth in statistical information and environmental observation allowed the creation of maps such as these. They follow the common convention of blue for cold, red for warm.

80 OVERLEAF The Enlightenment was a time when new phenomena were discovered and when new methods of mapping them were needed. This 1702 map by Edmond Halley uses the new cartographic concept of isolines to show variations in the Earth's magnetic field.

To the Right Honourable the LORDS COMMISSIONERS for executing the Office of LORD HIGH ADMIRAL of GREAT BRITAIN &c.

This Variation Chart is most Humbly Inscrib'd by their Lordships most Obedient Servt. Willm. Mountaine. F.R.S.

RUSSIA OR MUSCOVY

EUROPE

GERMANY

FRANCE

HUNGARY

POLAND

CASPIAN SEA

BLACK SEA

MEDITERRANEAN SEA

BARBARY

AFRICA

EGYPT

ARABIA PETRAEA

ARABIA DESERTA

ARABIA FELIX

THE RED SEA

TURKISH EMPIRE

EMPIRE of PERSIA

INDIA

INDOSTAN

GOLCONDA

BENGALA

BAY OF BENGALA

SIAM

PEGU

BAY OF TUNKING

EAST TARTARY

EMPIRE OF CHINA

AFRICA INTERIORA INCOGNITA

CONGO

ANGOLA

MOZAMBIQUE

ZOFALA

MONO MOTA PA

CAFFRARIA

HOTTENTOTS

MADAGASCAR OR ST LAURENCE

THE INDIAN SEA

MALDIVA ISLANDS

CEYLON

STREIGHT OF BABELMANDEL

West Variation

East Variation

THE INDIAN EASTERN OCEAN

DEGREES OF VARIATION WES...

graphically, particularly through maps, only developed in the first half of the nineteenth century.

Because most of the data being gathered had a spatial element, maps were an obvious means of visualizing it. Methods for the creation of thematic maps – maps focusing on one specific subject, as opposed to general maps depicting numerous landscape features together – were established as a matter of urgency as scientific and geographical discoveries were made. A particular challenge was that of mapping *quantitative* data – data with numerical or mathematical value, rather than data simply identifying the existence and location of a thing or event. Such techniques as *isolines* (lines joining points of equal value of a measured phenomena), *choropleth maps* (showing areas by colour or pattern according to the calculated density of phenomena within the areas) and *graduated symbols* (drawn in proportion

81 Based on the 1870 census of the United States, this choropleth map shows population density for defined areas, with population of the largest cities indicated by circles 'proportionate to population'. *Constitutional Population*, from *Statistical Atlas of the US*, Francis A. Walker, 1871.

MAP
SHOWING IN FIVE DEGREES OF DENSITY, THE DISTRIBUTION,
WITHIN THE TERRITORY OF THE UNITED STATES, OF THE
CONSTITUTIONAL POPULATION
(i. e., excluding Indians not taxed)
Compiled from the Returns of Population at the Ninth Census
OF THE UNITED STATES, 1870,
BY
FRANCIS A. WALKER.
To which is added a sketch of the principal
INDIAN RESERVATIONS AND RANGES
from information furnished by the Office of Indian Affairs
of date 1871.

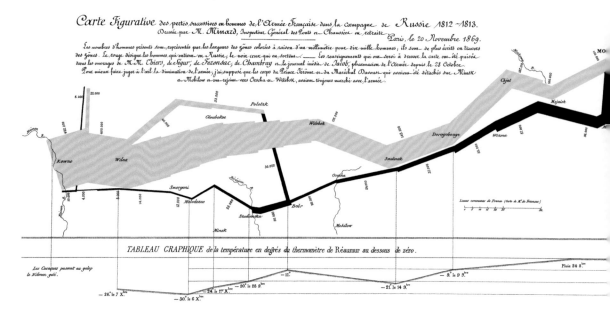

82 *Carte Figurative des pertes successives en hommes de l'Armée Française..., 1812–1813.* C.J. Minard, 1869. The thickness of Minard's lines represent the number of soldiers at different stages in Napoleon's Russian campaign of 1812 – a thick brown line for those entering Russia, thin black lines for the few returning.

to the statistical values they represent) were all developed at this time. In fact, most thematic mapping techniques in use today were fully developed by the late nineteenth century.

Several prominent scientists played a part in this. The English astronomer Edmond Halley first used isolines to portray variations in the Earth's magnetic field (lines known as *isogones*) as long ago as 1701, but the German geographer Alexander von Humboldt firmly established this method with his influential map of *isotherms* (lines joining points of equal temperature) in 1817. French mathematician Baron Charles Dupin created the first choropleth map (on the subject of education) in 1827, and several scientists/cartographers began to use graduated circles and lines to represent population and transport statistics. One particularly significant map (or what may now be called an 'infographic') was produced by the French civil engineer Charles Joseph Minard in 1869. It uses proportional lines to dramatically depict the depletion of Napoleon's troops during their Russian campaign of 1812 and their subsequent retreat from Moscow.

Cartography offers many tools to portray quantitative data. Point, line and area symbols (see p. 57) are able to reflect the data by variations in shape, size, colour and orientation to distinguish phenomena and their relative values. But part of the cartographer's task is to understand the data being used and to choose the most appropriate method to portray it. Choropleth maps, for example (fig. 81), generally divide data into classes, which are then represented by a sequence of colours, or gradations of a single colour – lighter colours commonly referring to lower values, darker colours to higher. But there are different methods of defining these classes, and if an inappropriate method is used the patterns shown on the map can change dramatically, telling an apparently different story. This effect may be inadvertent and innocent, but can also be used deliberately if the map-maker is following a particular agenda. Maps of election results or political campaigns are often guilty of using data classes or other statistical devices that suit one side of the story.

The huge amount of data available today, and mapping applications which allow users to create their own thematic maps from it (see p. 213), present great opportunities for thematic maps, which remain one of the most effective ways of presenting complicated information. There are dangers, however, if the data is misunderstood, processed and presented inappropriately or if a cartographer's personal (or corporate or political) motives are allowed to dominate the map-making process.

BREAKING THE CONVENTION

Socio-economic and environmental statistics are mappable in many ways, but one of the most dramatic methods is the use of *cartograms*. These are maps which distort a geographical area according to other variables such as population, wealth, tree cover or death rates. In population-based cartograms, China, India and Indonesia become greatly enlarged relative to other countries. Such 'value-by-area' maps can sometimes change our perceptions of the world.

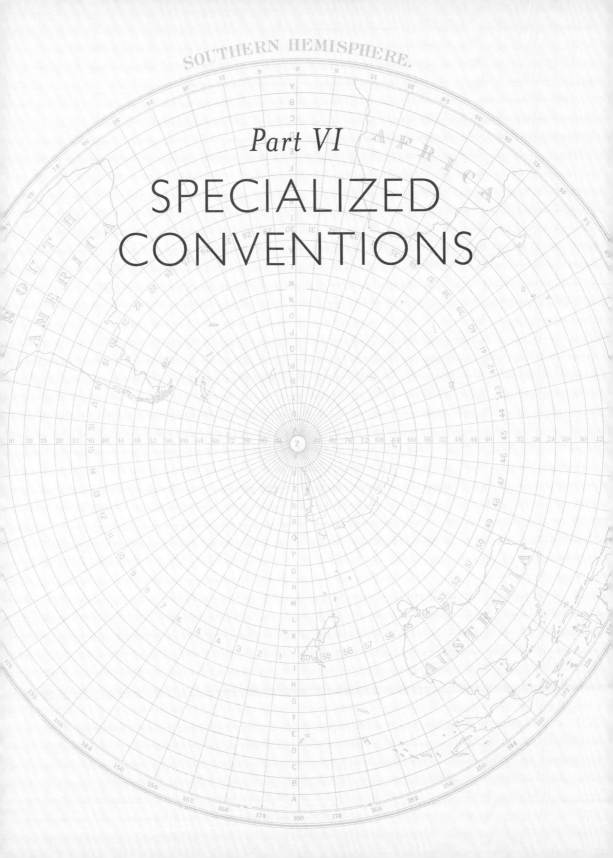

Part VI

SPECIALIZED CONVENTIONS

GEOLOGICAL MAPS
What lies beneath

No one would have thought that the building of Britain's canals in the late eighteenth and early nineteenth centuries would have such far-reaching implications for science and religion. But when William Smith, the engineer and surveyor of the Somersetshire Coal Canal, in 1793 explored cuttings made for the canal and descended into the mines that it would serve, he discovered things which challenged the status quo and changed forever our understanding of the Earth and how it was formed. In much of the world, religious doctrine of the time stated that the Earth was formed at a precise, very recent date, calculated from accounts in the Bible. That date, according to James Ussher, Bishop of Armagh, in 1658, was 4004 BCE. It was heretical to suggest anything else.

What Smith discovered was that rocks appeared to have been laid down in layers – or *strata* – which followed specific sequences. Also that the types of fossils found in each layer of rock could date the layers. These discoveries, and years spent examining rock outcrops across the country, allowed him to construct his hugely influential 1815 map *Delineation of the Strata of England and Wales* (fig. 83) and to gain the unofficial title of the 'Father of Geology'. He found that certain sequences of rock appeared to be 'upside down', with older rocks above younger ones. This suggested that

83 Extract of Somerset from William Smith's *Delineation of the Strata of England and Wales*, 1815. Smith's map revolutionized both thematic cartography and geology itself. The colours represent the rock types, with their intensity reflecting the rocks' relative depth.

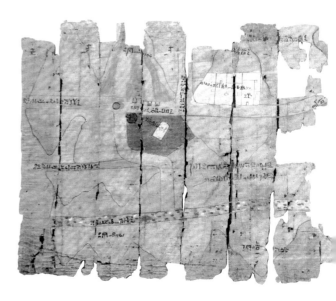

84 Early recognition of the value of natural resources is revealed in this Ancient Egyptian map. It shows a region where rocks for carving monuments were quarried, and where gold was mined. Part of the *Turin Papyrus, c.* 1150 BCE.

things were not quite as they seemed, and his descriptions and explanations of fossils led people to question widely held beliefs that evolution and extinction of species were impossible.

The driving forces behind Smith's work were the desire to exploit natural resources – in particular the coal and iron ore behind the Industrial Revolution – and the need to transport them efficiently, in his case by canal. The economic value of rocks and minerals has been behind much of the development of geological and mineralogical mapping over the centuries. Even one of the most famous maps from Ancient Egypt – known as the *Turin Papyrus* (fig. 84) and produced *c.* 1150 BCE – shows areas of gold-bearing rocks and the location of a gold-mining settlement. Interest in the Earth's resources was behind the development of mineralogical maps in the sixteenth century, particularly in Germany. It was eighteenth-century France, though, which was a focus for general geological mapping. Such mapping steadily grew in its sophistication as knowledge and understanding of geology increased. A mineralogical map of the Danube basin from 1726 by Count Luigi Ferdinando Marsigli used symbols to identify the locations of fourteen different minerals and ores. A little later, in 1746, Philippe Buache produced maps for the geologist Jean-Étienne Guettard which used around fifty different symbols. When Henry De la Beche – who was later to become the

first director of Britain's Ordnance Geological Survey in 1835 – worked on a colour scheme for his survey of Devon in 1832, he defined fourteen types of rock formation. The British Geological Survey (which succeeded the Ordnance Geological Survey) now recognizes 2,500 types.

The first use of colour to distinguish rock types was in 1774, but the use of colour and symbols was mastered in the early nineteenth century with Cuvier and Brongniart's map of Paris (1811) and Smith's map. The complexity of geology has always presented a challenge to cartographers, particularly when looked at globally. Rock types and formations vary between regions and continents, and universal conventions for mapping them have remained elusive. Some similarities exist between geological maps around the world, but there is no standard colour scheme. Different approaches have been taken to colour. Smith was among those who tried to approximate the colours of the rocks themselves in his map colours; others have used the spectrum – red–purple–violet–blue–green–yellow – to depict the relative ages of rocks, from older to younger. The first International Geological Congress in Paris in 1878 discussed the issue but was inconclusive. Subsequent Congresses in Bologna (1881) and Berlin (1885) devised what became known as the 'International Colour System', which was adopted to a degree but was found to be well suited to Europe but less so for mapping parts of North America and Australia. An alternative 'American Colour System' was

Pl. XIII.

EXPLANATION OF COLORS.

EOZOIC

CAMBRIAN and SILURIAN

DEVONIAN

CARBONIFEROUS and PERMIAN

TRIASSIC and JURASSIC

CRETACEOUS

TERTIARY

ALLUVIUM

VOLCANIC

GEOLOGICAL
OF THE
UNITED ST
COMPILED B
C.H. HITCHCOCK AND
from sources mentioned
1874.

Lith. by J. Bien N.Y.

developed and adopted by the US Geological Survey, but this is also limited in the areas to which it can be applied.

Symbols representing particular geological phenomena, such as fault lines and dip and strike (the angle and direction at which layers of rock underlie adjacent rocks), are more standardized, as is the convention of adding abbreviated typographic 'symbols' – sometimes made up of three or four letters – in order to help users distinguish between sometimes subtly different colours. But in geological mapping cartographers hoping to establish universal conventions appear to have met their match.

William Smith knew that his map needed to be understand-able – a critical principle for good map-making, whether strict con-ventions exist or not. It was widely

85 Geological maps are commonly found in national atlases – in this case the *Statistical Atlas of the United States*. The rock colours are overprinted on a simple background map showing primarily state boundaries and rivers. *Geological Map of the United States*, Hitchcock & Blake, 1874.

understood, and its importance –
not just for locating rock types and
resources but also in challenging
prevailing beliefs of the time and
influencing mapping practice –
cannot be overestimated.

BREAKING THE CONVENTION

Geologists are used to mapping remote
regions and to interpreting landscapes
from a distance. But staff at the United
States Geological Survey in 1961 went
a giant leap further by producing a
map of the Moon. Based on telescope
photographs of the Moon's surface,
their 'Generalized Photogeologic' map of
lunar rock types and landforms played
an important role in planning for the
first Moon landing in 1969.

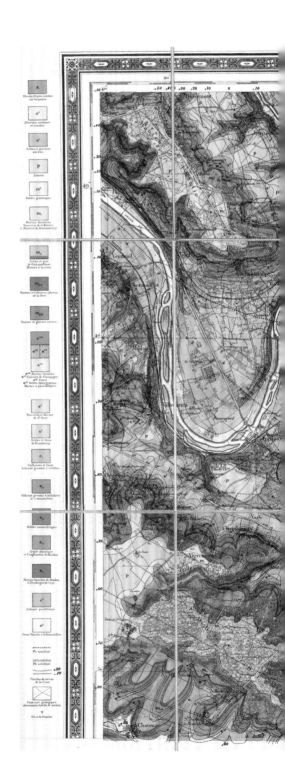

86 A complex and beautiful example of
well-executed geological mapping. This
is also an early example of the use of
typographic 'symbols' within the colours
to help distinguish rock types. *Paris et ses
environs. Carte géologique détaillée*, Ministère
des Travaux Publics, 1890.

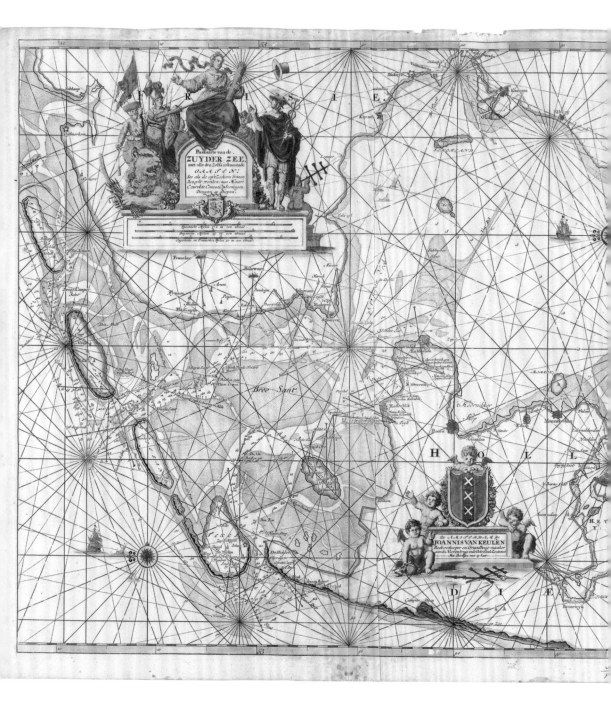

Paskaarte van de
ZUYDER ZEE,
met alle des Zelfs inkomende
G A A T E N:
Seo als die op't Zakerts binnen
Bezeylt worden: aan Minne-
Courebte Courses Subteraneca.
Droogtes, op Diepten.

t'AMSTERDAM By
JOANNIS VAN KEULEN
Boek-verkooper en Graadboog-maaker
aan de Nieuwenbrug in de Gekroonde Loodtens
Met Privilegie voor 15 Jaar

HYDROGRAPHIC CHARTS
Bon voyage

Between 1584 and 1585 a retired Dutch seaman produced a revolutionary collection of sea charts titled *Spieghel der Zeevaerdt*. So influential was this work by Lucas Janszoon Waghenaer that an English distortion of his name – 'Waggoner' – became established as the name for any collection of printed charts, whoever produced them.

Waghenaer's charts are beautifully engraved and highly detailed (fig. 88). They were the first to systematically depict features critical to navigation at sea, including soundings (see p. 99), detailed shoreline, terrestrial landmarks, buoys, daymarks, shoals and underwater rocks (represented by small crosses – a symbol still in use on charts today). These visionary products gave clear priority to information directly relevant to mariners, and established principles which continue to apply to modern nautical charts.

The Dutch dominated cartography and exploration at the time in which these products were made, and it was a time when developments in shipping and navigation were demanding (and generating) new, more accurate charts. Prior to this, charts had taken different forms, with very few common features or conventions between them. Products to aid coastal navigation date back to the fifth century BCE with the Ancient Greek *periplus* – lists of coastal features and ports rather than charts. Chinese river

87 The Dutch dominated the world of chart production for nearly two hundred years. This chart *Zuyder Zee*, 1681, by the prolific van Keulen publishing house is typical of the style and level of detail achieved during that period.

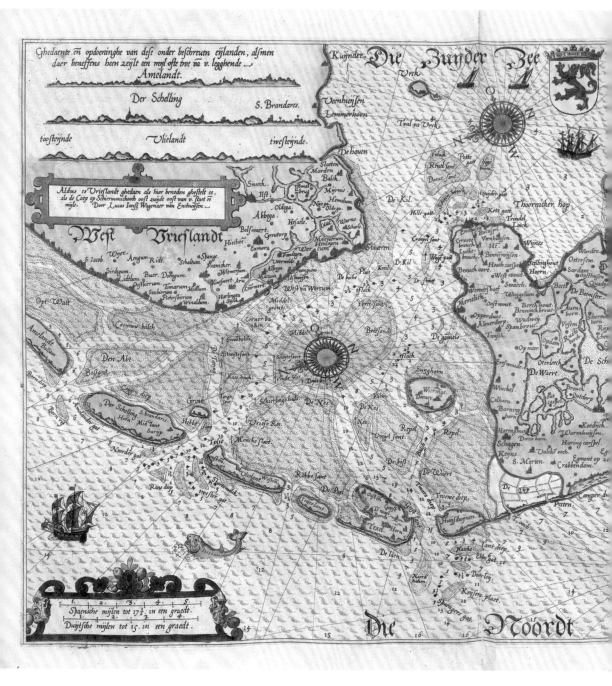

Ghedaente en opdoeninghe van dese onder beschreuen eijlanden, alsmen
daer beneffens heen zeijlt een myl oste twe vn v. legghende
Amelandt.

Der Schelling   S. Brandares.

toostreijnde   Vlielandt   tweestreijnde.

Aldus es Vrieslandt ghedaen als hier beneden gheftelt es,
als de Caep op Schirmonichooch oost zuijde oost van v. ftart en
myle. Deer Lucas Ianß Wagenaer van Enchuijsen.

West Vrieslandt

Die Zuijder Zee

Die Noordt

Spaensche mijlen tot 17½ in een graedt
Duijtsche mijlen tot 15 in een graedt

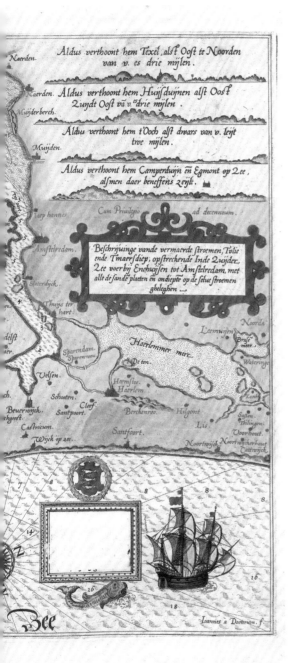

charts (indicating the importance of inland waterways in China) date from the sixth century CE and Arabic charts were produced in the tenth century. The most significant early navigational charts, however, are the portolan charts of western Europe and the Mediterranean, the oldest surviving example of which is the *Carte Pisane* of c. 1290 (fig. 89). A characteristic feature of these are *rhumb lines* (or *loxo-dromes* – lines of constant bearing, important for navigation), and they also established early conventions of showing important port names in red with lesser places in black.

Early charts were 'projection-less' or 'plane' charts – they paid no heed to the problems

88 Chart of the Zuider Zee from *Spieghel der zeevaerdt*, Lucas Janszoon Waghenaer, 1584. Waghenaer set the standard for nautical charts of the time, and also established conventions still followed today relating to symbols, navigational aids and depth information.

89 OVERLEAF The *Carte Pisane*, c. 1290, the oldest surviving portolan chart. The emphasis is clearly on the location of ports, and on rhumb lines to assist with navigation. The remarkable detail of the Mediterranean shows how well explored the region already was.

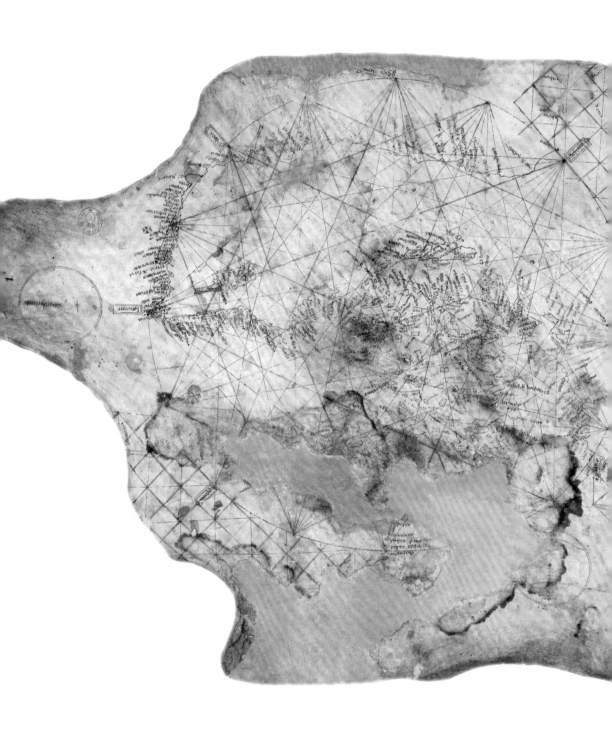

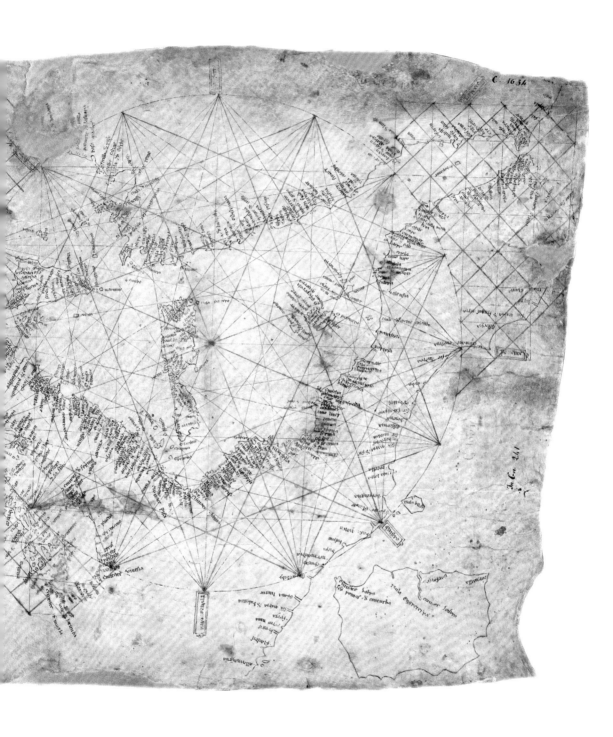

of representing the spherical Earth accurately in only two dimensions. Mercator's hugely influential map projection of 1569 (see fig. 10) changed things by bringing a much greater degree of accuracy while, significantly, allowing rhumb lines to be plotted as straight lines. Chart makers were quick to adopt his projection because of its benefits for navigation, and variations on it are still widely used.

Charts and sailing directions developed through the sixteenth and seventeenth centuries, notable publications being Portuguese *roteiros* (route descriptions) and coastal surveys such as that of Britain's coasts by Greenvile Collins in 1693. As survey techniques developed, expeditions by the colonial powers produced accurate charts of previously unmapped parts of the world. The Dutch continued to be a strong influence on charting and the van Keulen publishing house, which operated from 1678 to 1885, was a commercial forerunner to national hydrographic departments, which began to be established in the eighteenth century (France in 1720 and Britain in 1795 were the first). These departments were initially established for defence purposes, and it was through them that some of the symbols used on Waghenaer's charts, for example, became standardized. The need for mariners to be able to read charts and identify hazards and navigational aids in any part of the world was recognized.

Today, national hydrographic offices share the task of providing global chart coverage. The work is coordinated by the International Hydrographic Organization (IHO), whose hefty international chart specification – known as publication *S-4* – aims to ensure consistency between charts. To account for the vast range of information charts must carry – types of buoys, beacons, light sequences, hazards, and so on – it details hundreds of symbols. It also defines the colours to be used, including those which give charts their distinctive, conventional appearance – buff land, blue shallow water, green inter-tidal zone, white deep water. Chart makers have obligations to adhere to these standards, but can apply local variations where required.

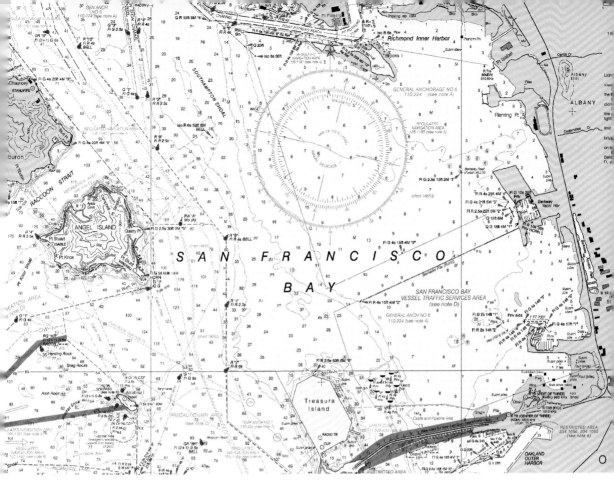

90 San Francisco, US National Oceanic and Atmospheric Administration, 2018. This extract illustrates the vast range of symbols and annotations on modern printed and digital hydrographic charts. Symbols and colours are now virtually standard across the world.

All this is to ensure compliance with the International Convention for the Safety of Life at Sea (SOLAS). Ships are obliged to carry approved charts – either paper charts or electronic navigational charts (ENCs), and must carry backups in case of loss or failure (a convention traceable back to 1354). They must also update their charts from regularly issued *Notices to Mariners*. Waghenaer recognized these issues and, having previously been a sailor for nearly thirty years, he understood their importance. Where life is at stake, mapping conventions become vital, and their agreement and implementation easier to attain.

MILITARY MAPS
Friend or foe?

There's a clue in the name. Ordnance Survey, Britain's national mapping agency, had its origins in the Board of Ordnance – the government agency responsible for military supplies and munitions. Following the Jacobite rebellion in Scotland in 1745, the Board recognized the need for detailed mapping to help keep things under control. In response, between 1747 and 1755, William Roy and his team produced a series of beautiful maps of Scotland showing relief, roads, settlements, drainage and even field patterns (fig. 91). When priorities changed with the start of the Seven Years War with France, Roy was posted to England, where he died in 1790, just a year before the founding of the Trigonometrical Survey, based in the Tower of London, which became Ordnance Survey.

Demands for military maps, on land, at sea or in the air, stem from perceived internal and external threats, and the need for governments to protect their citizens. This requirement goes back centuries and most maps throughout history have been of potential military use. The earliest truly military maps were found in Changsha, China, and date back to the second century BCE. One of these maps, printed on silk, shows water features, positions of army units, command posts and watchtowers, as well as settlements with the number of inhabitants. Information judged to be of military

91 Map of Loch Naver area, *Military Survey of Scotland*, William Roy, 1747–55. Roy's systematic survey of Scotland, with its detail of roads, drainage, settlements and relief, was an important influence on the establishment of Britain's Ordnance Survey.

use is shown in red. The map also shows what is now referred to as 'goings' information, or terrain analysis – details of how easy it is to travel, or 'go', through the area. River crossing points and roads are shown, giving a foretaste of twentieth-century Russian military mapping in particular, which shows the width and strength of bridges, road surfaces and the spacing of trees in wooded areas.

92 Mapping fortifications has long been an important part of planning military operations. This fort plan from North America details the function of individual buildings and includes plans for a new dock. *Plan of Annapolis Royal in Nova Scotia*, c.1744.

93 Trench maps were critical during the First World War. Enemy trenches are shown in red, with specific targets shown as circled numbers. The grid allows accurate targeting and coordination of operations. Extract from *France, 1:10,000 Series GSGS 2742, Fonquevillers*, War Office, 1916.

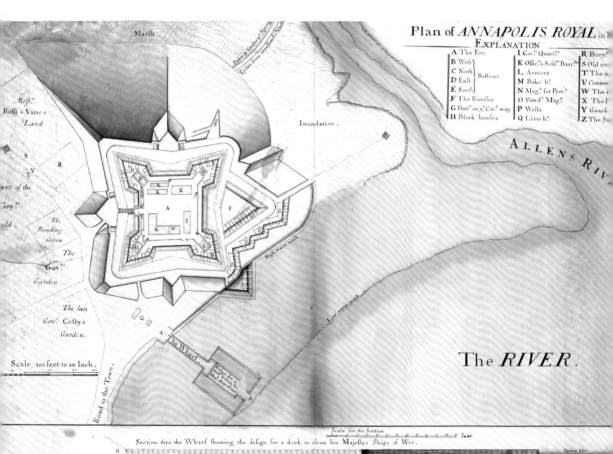

Plan of *ANNAPOLIS ROYAL* in

EXPLANATION

A The Fort	I Gov.ʳ Quart.ʳˢ	R Bury.ᵍ
B Weſt	K Offic.ʳˢ Sold.ʳ Barr.ˢ	S Old ro
C North	L Armory	T The n
D Eaſt	M Bake h.ˢ	V Commu
E South	N Mag.ˢ for Prov.ˢ	W The r
F The Ravelin	O Powd.ʳ Mag.ˢ	X The l
G Batt.ⁿ on y.ᵉ Cov.ᵈ way	P Wells	Y Guard
H Block houſes	Q Lime h.ˢ	Z The S

Scale, 100 feet to an Inch.

Section thro the Wharf ſhewing the deſign for a dock to clean his Majeſtys Ships of War.

Scale for the Section feet.

The Romans made maps of their own forts and encampments, but it wasn't until the fifteenth century that maps specifically for military campaigns began to be produced. A notable example from 1469 is an Italian map for the first Venetian–Turkish war. Henry VIII mapped Britain's coastal fortifications from the 1530s in manuscript form (it was always a good idea not to produce multiple copies of sensitive military documents), at a time when cartography was becoming a task for surveyors rather than artists. Regional military maps of the seventeenth century were the forerunners of national topographic surveys, which developed through the next two centuries. Detailed knowledge of terrain is critical to military operations, and authorities around the world quickly recognized the value of detailed, contoured topographic maps. These became, and remain, the basis for modern military mapping, and in some countries are still themselves treated as sensitive, classified documents.

The First World War was the first major conflict in which maps played a crucial role – over 34 million maps were produced by the British during the course of the war. With artillery playing an important part, targeting information had to be accurately plotted, as did the alignment of trenches and troop positions. For the first time, aerial reconnaissance helped with this, providing vital intelligence for tactical maps and those showing the status of operations. Because of what was at stake, accuracy and clarity were of the utmost importance (literally a matter of life or death); it was during this time that conventions such as the use of red to represent enemy forces and blue for allies began to develop (figs 93 & 94).

Military mapping serves many purposes – planning, operations, reconnaissance, training – and distinctive symbols, such as those in NATO's *Joint Military Symbology*, have developed. These ensure consistent depiction

94 Aerial photography played a crucial role in the compilation of large-scale topographic maps of France in preparation for D-Day. It also provided critical information on enemy positions which was overprinted on the maps. Extract from *France 1:25,000, Defences, BIGOT*, GSGS, 1944.

of troop affiliations, units –
infantry (diagonal cross), medical
(horizontal cross) and artillery
(solid circle), for example – and
location of equipment and instal-
lations. Standard topographic maps
usually form the basis for military
maps with the supplementary
information overprinted – symbols,
tactical information and grids to
help determine and communicate
position more accurately (see p. 29).
Also commonly included in their
legends (see p. 41) are series, sheet and edition details for each map, to
ensure that allied troops are all referring to precisely the same document in
planning and carrying out manoeuvres.

BREAKING THE CONVENTION

During the Second World War the
British intelligence service created a
new branch – MI9 – to help servicemen
evade capture and escape if taken
prisoner. Part of their task was to
distribute maps for potential escapees.
These were printed on silk, so that
they could be packed small and hidden
within such things as board games and
vinyl records - items which could be
legitimately taken into prisoner-of-war
camps.

The technologies of war have often led cartographic developments.
Increasing precision in weapons systems has made new demands on maps
and geographic information. Just as map data now forms the core of the
activities of Britain's Ordnance Survey, so has geographic information
in its broadest sense – digital maps, aerial and satellite imagery, digital
terrain models, GPS – become the lifeblood of modern military activities.
Printed maps may not be so common on the battlefield but the principles
and conventions behind them are well embedded in modern weapons and
command-and-control systems.

GLOBAL MAPPING
Let's stick together

It seemed such a good idea. Each country would map its own territory at a scale of 1:1,000,000, to an agreed specification, following agreed conventions, on a single map projection and with precisely defined sheet lines. It would be an exercise in international cooperation for the common good, and create for the first time a consistent map of the whole world. The map would be invaluable for administrative, planning and development purposes and as a base map for thematic maps of topical issues. Such was the vision of Albrecht Penck, a distinguished German geographer, when in 1891 he proposed the *International Map of the World* (IMW) to the Fifth International Geographical Congress in Berne. What could possibly go wrong?

Quite a lot, as it happened. The idea was supported by many countries but the fact that it took twenty-two years for a map specification to be finally agreed was perhaps an indication of how torturous the project would prove to be, and how it would epitomize the difficulties of establishing universal cartographic conventions. Many maps were quickly produced under the project – most notably in South America – but only a fraction of these adhered strictly to the published specification. Several factors seemed to conspire against it. No sooner had the specification been agreed, and a Central Bureau for IMW been established in 1913, than the First World War broke out. The two world wars gave contributing countries quite different priorities – mapping was needed at larger scales, and maps and

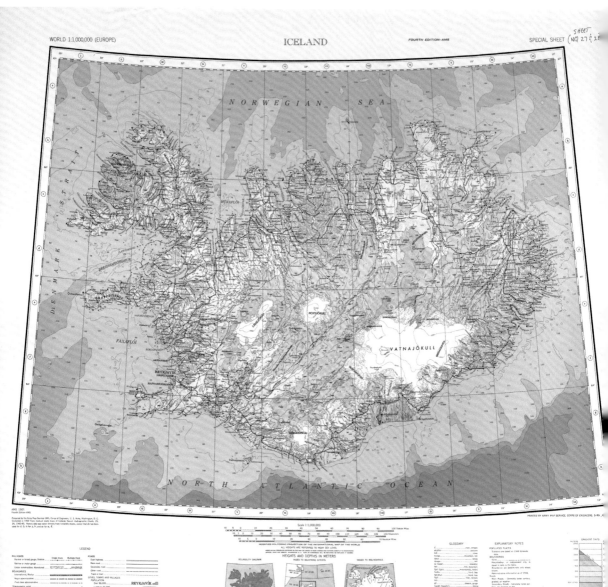

Scale 1:1,000,000

MODIFIED POLYCONIC PROJECTION OF THE INTERNATIONAL MAP OF THE WORLD

HEIGHTS AND DEPTHS IN METERS
ALL HEIGHTS ARE REFERRED TO MEAN SEA LEVEL

LEGEND

GLOSSARY

EXPLANATORY NOTES

RELIABILITY DIAGRAM

INDEX TO ADJOINING SHEETS

INDEX TO BOUNDARIES

ICEL

geographical information became highly sensitive. The sharing of information vital for the success of IMW was immediately challenged. Although maps at the same scale were produced during the war years, they served quite different purposes to the very general-purpose IMW and did not follow the IMW scheme.

It was also unfortunate for IMW that air travel developed so rapidly between the wars. For the first time, aeronautical charts were needed to ensure safe travel. The IMW specification had not anticipated this and did not meet the requirement. During the Second World War the US Coast and Geodetic Survey developed what became known as the *World Aeronautical Chart* (WAC), at the same scale as IMW. The urgency of producing these charts worldwide is reflected in the fact that its specification gained international agreement almost immediately in 1944. Conventions have always been easier to establish when safety is at stake, and the obvious purpose and value of WAC was something IMW clearly lacked. WAC used a map projection more suitable to navigation, had more flexible sheet lines than IMW's strict schema (fig. 96) and emphasized different content. It was unrealistic for countries to produce and maintain two map series at the same scale, and the production of WAC took priority after the Second World War and worldwide coverage was achieved relatively quickly.

The value and aims of IMW were still widely recognized, however, and despite its Central Bureau having suffered bomb damage during the Second World War, which destroyed its records and most of its mapping material, the project continued under the auspices of the United Nations in New York. But geopolitical interests, defence requirements, national self-interest, demands for flexibility in the map's specification and a tendency for nations

95 *Iceland, IMW Sheet NQ-27/28*, US Army Map Service, 1952. A wide range of the IMW's symbols for both land and sea are shown on this sheet. The layer colouring for relief and bathymetry is characteristic of the series.

96 OVERLEAF *Index diagram of sheets for the International map on the scale of 1:1,000,000*, Geographical Section, General Staff, 1909. This index to the IMW shows the ambitious extent of the project. The numbering system for the 2,500 sheets is shown in red.

NORTHERN HEMISPHERE.

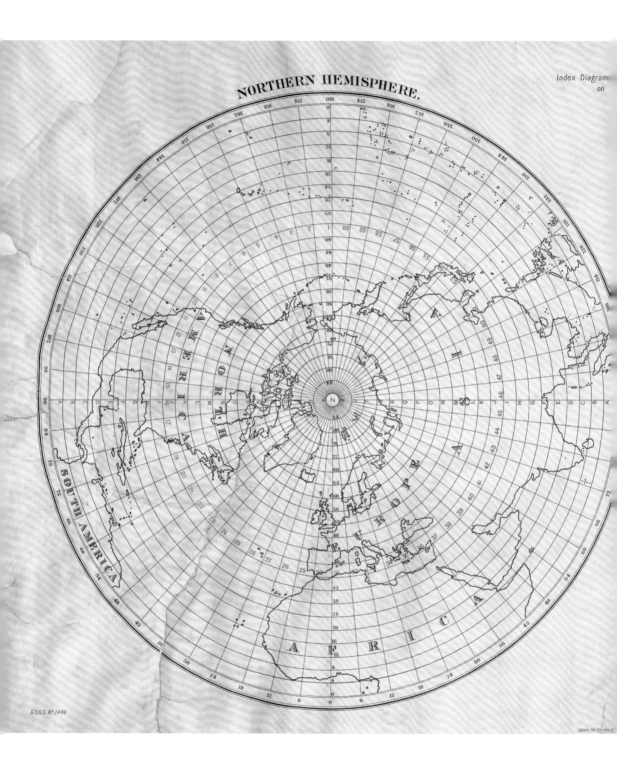

Agents for the sale of

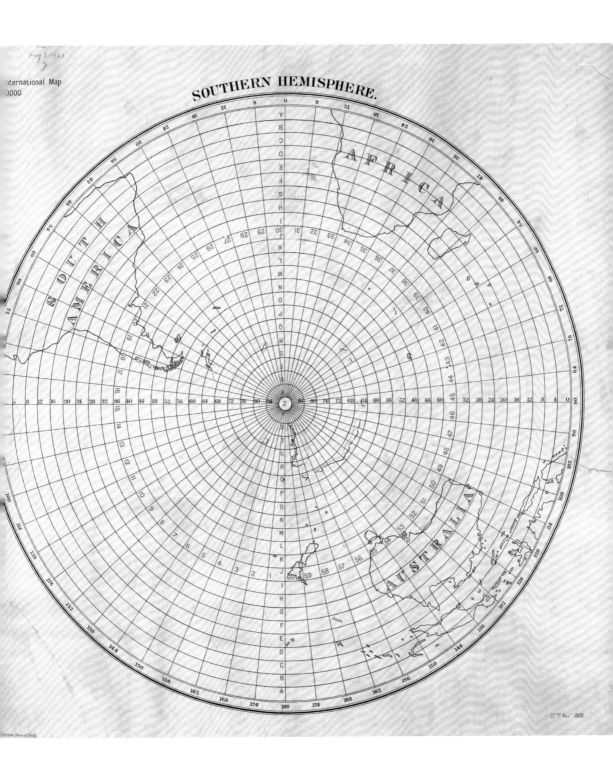

to want to map other countries' territory (something not allowed under IMW 'rules') continued to plague the project. Changes to the map agreed in 1962 came too late, and it soon became clear that it wasn't sustainable, so IMW was discontinued in 1989. It had by then achieved some success, having produced over 700 of the nearly 1,000 sheets needed to cover the Earth's land areas. However these did not always strictly follow the IMW specification, and were even by then well out of date.

The IMW tried to use universally applicable conventions – a single projection; fixed sheet sizes covering 6° longitude by 4° latitude; a standard vertical interval for contours; defined map symbols; place names in the Latin alphabet in anglicized form. But with so many agencies involved and regional and national interests to account for – Egypt, for example, objected to green being used to represent relief layers (see p. 117) because it misrepresented its arid terrain – worldwide agreement was difficult to achieve.

Today, although their aims and style are quite different from IMW, online mapping applications such as Google Maps (see p. 207) provide worldwide uniformity and consistent base maps. They can also be updated almost instantaneously and easily combined with a multitude of global geographic datasets now available. Perhaps Penck would have appreciated their freedom from the need to get agreement and action from dozens of national mapping agencies with their own mapping agendas.

97 This map covers a low-lying area, with dark green representing land below sea level. It includes symbols for features associated with this type of landscape – salt flats, sand, marshland and areas subject to inundation. Extract from *Astrakhan' (East)*, *IMW Sheet NL 39*, US Defense Mapping Agency, 1979.

Part VII

POST-CONVENTION MAPPING

Crocodili.

Dactili.

Arena.

DIFFERENT PERSPECTIVES

Picture this

The Frenchman Gaspard-Félix Tournachon, known by his nickname Nadar, was a caricaturist and photographer with a fascination for manned flight. In 1858 he became the first person to take photographs from a hot-air balloon – quite an achievement, considering the photographic processes of the time meant that he had to effectively carry his darkroom and processing equipment with him in the balloon's basket. He was also the first person to take photographs below ground – in the catacombs of Paris – using artificial light. He clearly liked to see things from different perspectives.

Nadar allowed people to see views from above that could previously only have been imagined. And yet maps drawn as bird's-eye views, or panoramas, had already been popular for nearly three centuries. Panoramic perspective maps of cities and towns were very popular in late-fifteenth-century Italy in particular; despite their views only being imagined, their level of detail was astonishing. The oldest surviving example, and perhaps the most remarkable, is the large view of Venice created by Jacopo de' Barbari in 1500 – a six-sheet woodcut print measuring 135 × 282 cm (fig. 99). The trend spread to the rest of Europe, notably Germany, and numerous town plans were drawn in this style, with buildings and features represented as three-dimensional drawings.

98 Extract from *Cairo*, from *Civitates orbis terrarum*, Braun and Hogenberg, 1572. This is one of over 500 perspective views in Braun and Hogenberg's 'Cities of the World', produced between 1572 and 1617. Each map was accompanied by a description of the city.

The portrayal of hills and mountains in elevation (side-on) or perspective (bird's-eye) view is found on some of the earliest maps, and perspective drawings of buildings to represent settlements have appeared on maps since Ancient Greek and Roman times (see p. 62). Both methods remained popular through medieval times – a period which also saw the creation of the earliest regional panoramic maps, of Asti and Alba in Italy, in 1291. Three-dimensional representation of cities in particular then

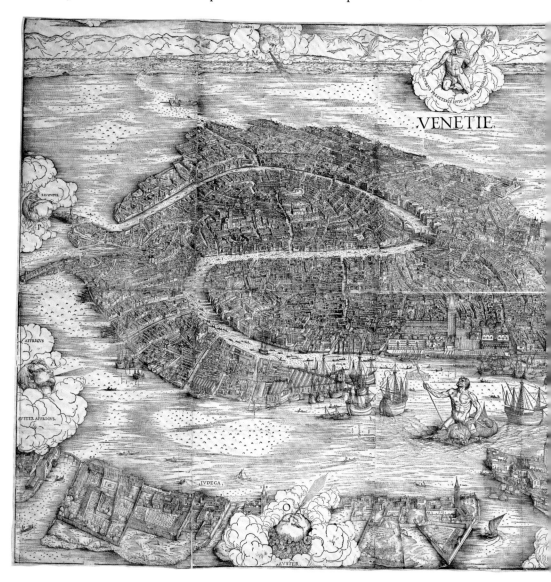

99 *Venice*, Jacopo de' Barbari, 1500. Without aerial photographs for reference, this bird's-eye view is remarkable in its level of detail. It not only shows the buildings of Venice, but also gives an impression of life on the canals.

100 OVERLEAF *A Balloon View of London*, Edward Stanford, 1859. Drawn as if from a hot-air balloon, this detailed, engraved, hand-coloured map looks south over London, and distorts perspective to allow more distant areas to be shown.

THE RIVER

THE RIVER TH

101 *Yellowstone National Park*, Heinrich Berann, 1991. Berann perfected the technique of panoramic maps, creating over 500 throughout his career. This is one of a series he produced to promote the dramatic landscapes of the national parks of the USA.

steadily grew in sophistication through Barbari's time into the eighteenth and nineteenth centuries when the technique benefited from the invention of the hot-air balloon in 1783. There remains a great fascination for maps using the artistic three-dimensional approach.

Because bird's-eye views and panoramas fall somewhere between cartography and art, normal cartographic conventions aren't generally followed. Although the most compelling bird's-eye city views consist of three-dimensional drawings of buildings placed in their correct planimetric position, traditional ideas of scale, map projection, orientation and symbology commonly don't apply to this type of map. Scale is adjustable to give features extra prominence, visual distortions are applied to ensure features are not hidden behind other features or don't disappear over the horizon, and north is not always at the top – panoramic maps of the Cairngorms National Park in Scotland, for example, were created from five different viewpoints, looking into the Park from each of its five main entrance points.

Such artistic licence is often used to dramatic effect in modern panoramas of mountain regions. The master of landscape panoramas was the Austrian Heinrich Berann, who from 1934 to 1996 created well-known panoramic maps of the Alps and of some of the USA's national parks (fig. 101). By including the sky, which almost by definition never normally appears on maps, these products certainly bring a new dimension to mapping. Berann also produced the artwork for the iconic three-dimensional global image of the sea floor produced by Marie Tharp and Bruce Heezen in 1977.

Nadar would have been delighted with the development of aerial photography and satellite imagery, which have become vital tools in representing the third dimension. Highly accurate measurements are now made from aerial photographs and from other forms of remote sensing – for example, laser-based light detection and ranging (LIDAR) – which allow the creation of precise, photorealistic digital models of terrain (both terrestrial and undersea) and buildings. Visualizations of such data are now common through web mapping applications such as Google Maps. Nadar would no doubt also be pleased that similar tools are used by geologists and archaeologists to visualize underground features. 'Mapping' is now a term not restricted to two-dimensional planimetric views, but one which is also commonly applied to graphic visualizations of highly complex three-dimensional data.

DIGITAL &
ONLINE MAPPING

In our hands

It's hard to imagine life without Google. Originally formed in 1998, it now impacts on many aspects of our lives and has billions of users across the world. Google's philosophy is defined by their 'Ten things we know to be true', and one of these is that 'It's best to do one thing really, really well.' So what do they do? In their own words, 'We do search.' Google's success was built on their focus on searching the World Wide Web and delivering relevant results to users. When, in 2005, they began to base searches on location, using maps both to define searches and to display search results, the Google Maps and Google Earth applications set them apart and helped lead to the ubiquity of online maps today.

Maps were always around, but were not always easily found, and by being on paper they were 'fixed' objects, representing a moment in time. Having become digital, and by being able to capitalize on other technologies such as the Global Positioning System (GPS), mobile tele-communications and smartphones, maps are now in our cars (as satnav) and in our hands everywhere, all the time, and are being constantly (and almost instantaneously) updated.

Geographic information systems (GIS) have been critical in these developments. GIS date back to the 1960s, but became established in 1982

102 Sample of *MasterMap* data, Ordnance Survey, 2018. MasterMap is the Ordnance Survey's most detailed digital map of Great Britain. It forms a seamless, regularly updated database, widely used by central government, local authorities and utility companies.

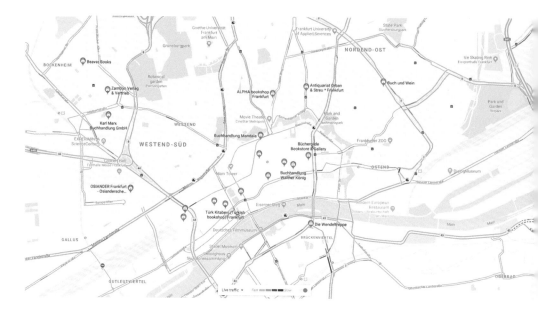

103 Map of central Frankfurt, Google Maps, 2018. This screenshot shows the locations of bookshops and the status of traffic, which is constantly updated. Each bookshop 'drop-pin' is a live link to the shop's address and opening times.

104 Screenshots from *windy.com* online weather maps. These examples show wind, wave height and temperature. Based on several climate models, and with constant updates, this application allows users to display worldwide, animated maps of numerous climate phenomena with detailed forecasts.

with the release of ARC/INFO by Environmental Systems Research Institute (ESRI), which has become the most widely used GIS software in the world. GIS allow the gathering, processing, analysis and presentation of any data which has a geographical element. Features definable by their location – some would say this is anything and everything – may be treated as separate layers of information for display on a map. Relationships between layers are able to be analysed, and each layer can have its own distinct 'attributes'. A road, for example, may show as a simple line marking its route, but may also be given attributes defining such details as the nature of its surface, its suitability for certain types of vehicle, which authority has maintenance responsibility for it or traffic levels at different times of day.

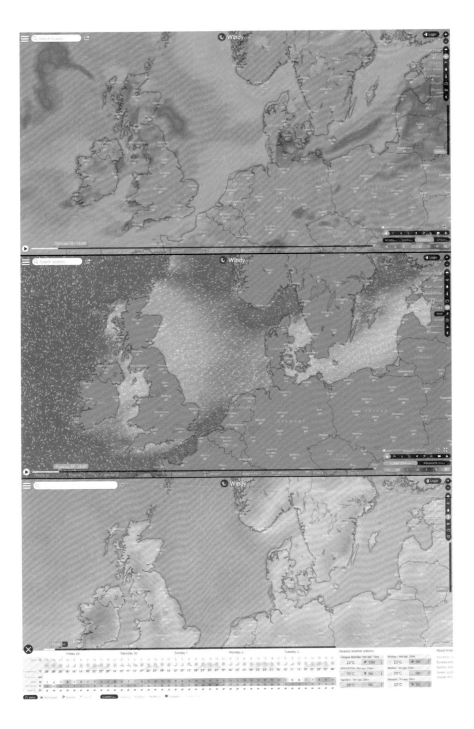

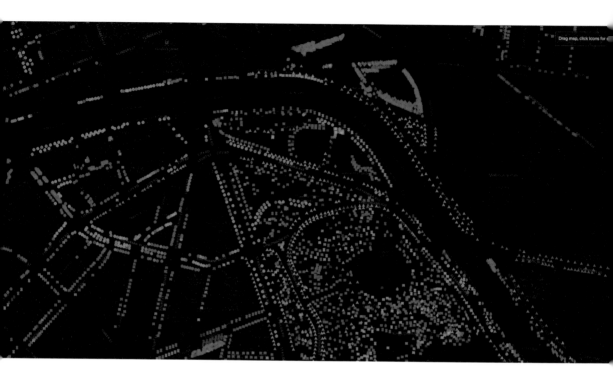

Drag map, click icons for

105 *Urban Forest Visual*, City of Melbourne, 2018. Extract from Melbourne's online mapping application, which locates the city's 70,000 trees and classifies them by general status, genus and 'useful life expectancy' – a measure of their health.

Such attributes and the way they are handled now influence the content of the maps we see on our screens.

GIS aren't maps as such, but maps are the most common way in which GIS data is presented or visualized. When data is displayed as a map, normal cartographic conventions may, or may not, apply. With the multi-disciplinary nature of GIS – it is widely used in environmental sciences and archaeology, for example – many non-cartographers, who may not be familiar with these conventions, are now producing maps. Fundamental conventions still prevail within GIS and the online mapping systems which rely on them – map projections, latitude and longitude, generalization (in relation to levels of detail at different zoom levels) and point, line and area symbols — but maps from GIS may not adhere to traditional style conventions as closely as they perhaps should.

Digital and online mapping conventions now relate more to functionality. Google Maps did not set out to be deliberately disruptive in relation to cartography, but rather followed technological developments, user needs and the extent of available data. While some aspects of their maps appear fairly traditional, new conventions have emerged. Zoom, pan, swipe, click, tap and linking to other sources of data are now as crucial to online maps as to any smartphone app. Similarly, a single tap can centre the map on the user's location, and significant places marked by a drop-pin, or even virtually 'visited' by placing a stick figure on the location to see photographs of the place. Users can interact with maps in completely different ways (see p. 213), with maps taking account of individual preferences and showing information directly relevant to the user's location. The ability to search has also become a convention — searching for routes, places, nearby restaurants, even for where your friends are, are now everyday functions which allow people to engage with, appreciate and 'do search' with maps in new ways.

(see p. 213)

BREAKING THE CONVENTION

The Internet, which has made the global distribution of maps easy, and has been behind the huge growth in the use of maps, has itself been mapped. Geographical location is not particularly important in the way digital data moves around the world, giving cartographers freedom to devise new and sometimes beautiful - an example from 2003 by the Opte Project is in New York's Museum of Modern Art - ways of mapping Internet traffic and connectivity.

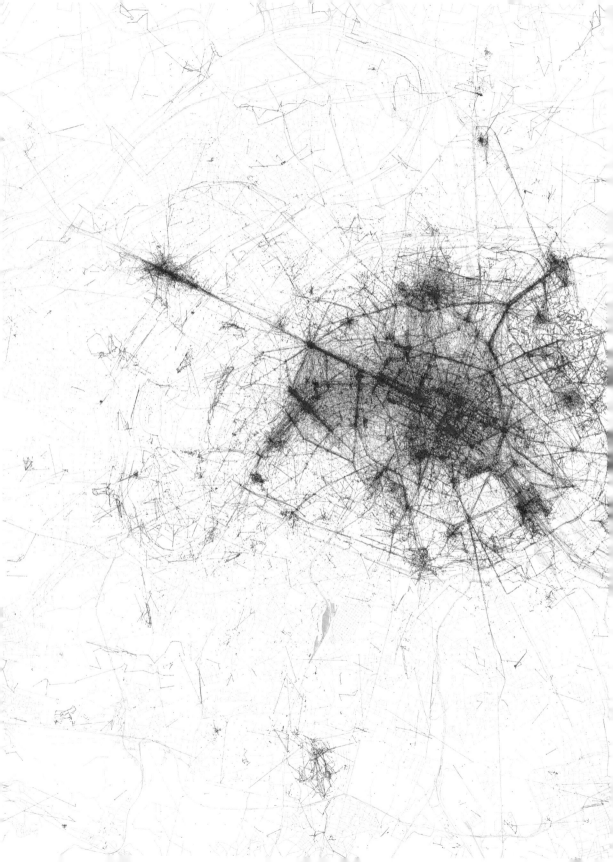

INTERACTIVE MAPS & DEMOCRATIZATION
MIY — Map it yourself

In a speech to the Royal Geographical Society in April 1914 Rudyard Kipling said, 'as soon as men begin to talk about anything that really matters, someone has to go and get the atlas.'[7] He appreciated that geographical knowledge was important in understanding current affairs and wanted to engage with maps. Where would he go today? Printed atlases still exist, but vast amounts of geographical information are now at our fingertips thanks to digital maps, the Internet and smartphones. Would these have given Kipling a greater understanding of what is going on in the world?

The way we interact with maps has changed immeasurably since then. Online maps now incorporate high levels of interactivity, allowing and encouraging us to adapt them to our own preferences and to contribute directly to the maps we see on-screen. This involvement of users, and the movement away from maps being created by professional cartographers, trusted atlas publishers or national mapping agencies, is often referred to as the 'democratization' of cartography. People can now, for better or worse, express themselves and their causes through the creation of maps and disseminate them instantly across the Internet.

106 Map of Paris by Eric Fischer, 2010. This map was created by plotting the locations of photographs taken by locals (blue), tourists (red) or either (yellow). The photographers have made an inadvertent contribution to the map by geotagging their photographs.

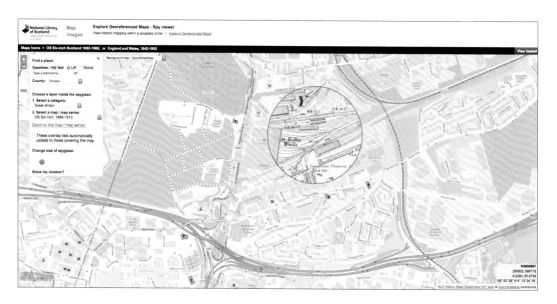

107 The National Library of Scotland mapping application provides access to thousands of digitized historical maps, allowing comparisons between map series and between maps of different dates. Here a 'spyglass' reveals an 1896 map behind a modern map of Glasgow.

Online mapping applications allow users to change the appearance of maps, contribute information to be shown on them and adjust their functionality. And there appear to be no rules – cartographic conventions may not feature highly, if at all, in a user's thinking. But this freedom to be involved – the ability to become a map-maker – is itself perhaps an important new mapping convention.

In online mapping applications you can change colours, switch layers of information on or off, or change a map's orientation – all to present the map in the way you think best for your purpose. Obvious dangers may follow – statistics could be mapped in a way that distorts their true meaning, inconvenient data which doesn't fit the author's purpose could be omitted and poor use of colour may mislead. But, at the same time, these user-generated maps might result in greater understanding, and some adaptations (careful choice of colours to allow for users with colour blindness, for example) can make a map instantly more accessible.

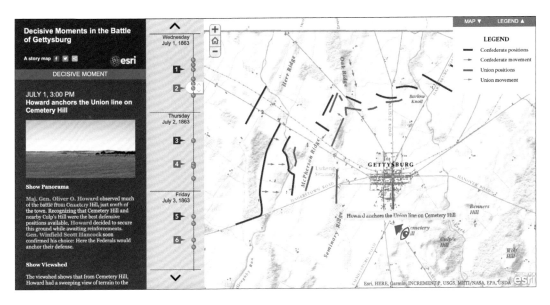

108 Part of a 'Story Map' about the Battle of Gettysburg of 1863. ESRI's Story Maps allow users to create their own maps and to combine these with text and images to form interactive stories which can be delivered online.

Information to enhance a map's content can be contributed deliberately or involuntarily. Applications now allow the marking of significant locations by drop-pins, with links added to a user's own content such as geo-tagged photographs (fig. 106). Increasing amounts of 'open source' (free to use) data is now available – which it is possible to incorporate into online maps without the need for specialist GIS software or skills.

What is perhaps less obvious, potentially sinister, is that people may be contributing information to online maps without realizing it. Details of our activities, movements and online searches is readily gleaned by Internet providers, mobile phone operators and search engines, and incorporated into the data behind online maps. Updates to traffic data, for example, are gathered in this way, but so too is data which people may prefer to keep to themselves.

Creators of online mapping applications could be expected to be against this level of interaction, wishing to maintain editorial control over their maps. But it is they who have made it possible. And by providing application programming interfaces (APIs) they even allow developers to change how the applications themselves work – datasets are combined, changes made to the base map, overlays added and maps embedded into websites.

Epitomizing this democratized cartography is OpenStreetMap (OSM). Begun in 2004 this collaborative, community-based mapping project now provides an open-source global map which has been built up by over a million contributors, volunteering their time and skills (fig. 109). And through its Humanitarian OpenStreetMap Team (HOT), it has responded to natural disasters – in Kipling's terms, things that really matter – around the world by creating much-needed mapping for relief work. By using online applications, people can quickly create maps in a crisis situation, providing a crucial tool for rescue operations.

BREAKING THE CONVENTION

In the 1760s the English cartographer John Spilsbury introduced a new way of interacting with maps – the jigsaw puzzle. For twenty years or so, maps were the only images to appear on what were originally known as 'dissected' puzzles. The earliest examples were for educational purposes. Pieces were simpler than in modern map puzzles available today, and were cut along coastlines, country boundaries and lines of latitude and longitude.

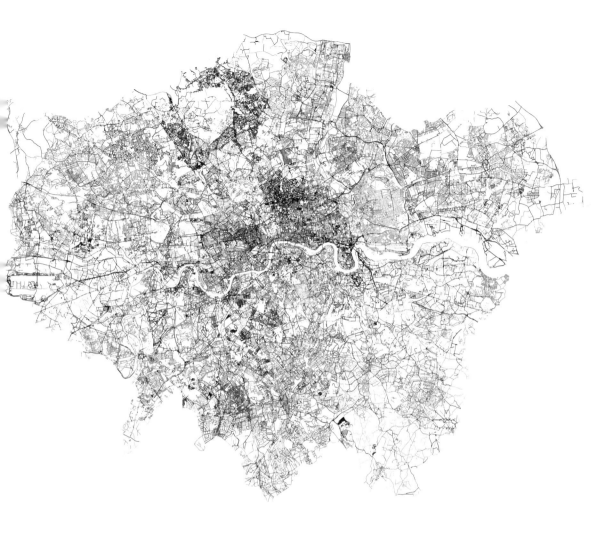

109 Map of contributors to *OpenStreetMap*,
Oliver O'Brien, 2014. The colours show the
features mapped in London by twenty-five
major contributors to the collaborative
mapping project, which has had over a
million contributors worldwide.

NOTES

1. See also https://en.wikipedia.org/wiki/ On_Exactitude_in_Science. Borges and Carroll quotations from 3stages.org/c/ gq.cgi?first=QAMAP.
2. Ibid.
3. John Smith, *The Art of Painting in Oyl*, Printed for S. Crouch, 1701.
4. Mark Monmonier, *How to Lie with Maps*, 3rd edn, Uiniversity of Chicago Press, Chicago, 2018, p. 118.
5. Edouard Imhof, *Cartographic Relief Representation*, ESRI Press, Redlands CA, 2007; original German edition, *Kartographische Geländedarstellung*, Walter de Gruyter, Berlin, 1965.
6. A. Robinson, R. Sale and J. Morrison, *Elements of Cartography*, Wiley, New York, 1978, p. 321.
7. Rudyard Kipling, 'Some Aspects of Travel', speech delivered to the Royal Geographical Society, 14 February 1914; published in *The Times*, 18 February 1914, and in *Journal of the Geographical Society*, April 1914. Published online at www. telelib.com/authors/K/KiplingRudyard/ prose/BookOfWords/aspectstravel.html.

FURTHER READING

Ashworth, M., and P. Parker (eds), *History of the World in Maps: The Rise and Fall of Empires, Countries and Cities*, Times Books, Glasgow, 2015.

Barber, P., (ed.), *The Map Book*, Weidenfeld & Nicolson, London, 2005.

Brotton, J., *Great Maps: The World's Masterpieces Explored and Explained*, Dorling Kindersley, London, 2014.

Carlucci, A., and P. Barber (eds) *Lie of the Land: The Secret Life of Maps*, British Library, London, 2001.

Cheshire, J., and O. Uberti, *London: The Information Capital*, Penguin, London, 2014.

Crane, N., *Mercator: The Man Who Mapped the Planet*, Phoenix, London, 2003.

Desimini, J., and C. Waldheim, *Cartographic Grounds: Projecting the Landscape Imaginary*, Princeton Architectural Press, New York, 2016.

Dorling, D., and D. Fairbairn, *Mapping: Ways of Representing the World*, Addison Wesley Longman, Harlow, 1997.

Foxell, S., *Mapping England*, Black Dog, London, 2008.

Garfield, S., *On the Map: Why the World Looks the Way It Does*, Profile, London, 2013.

Hall, D., (ed) *Treasures from the Map Room*, Bodleian Library, Oxford, 2016.

Imhof, E., *Cartographic Relief Presentation*, ESRI Press, Redlands CA, 2007.

Keates, J.S., *Cartographic Design and Production*, Longman, London, 1980.

Mitchell, R., and A. Janes, *Maps: Their Untold Stories: Map Treasures from The National Archives*, Bloomsbury, London, 2014.

Monmonier, M., *How to Lie With Maps*, University of Chicago Press, Chicago, 2018.

Robinson, A., R. Sale and J. Morrison, *Elements of Cartography*, Wiley, New York, 1978.

Seed, P., *The Oxford Map Companion: One Hundred Sources in World History*, Oxford University Press, New York, 2014.

Thrower, N.J.W., *Maps and Man: An Examination of Cartography in Relation to Culture and Civilization*, Prentice-Hall, Englewood Cliffs NJ, 1972.

Wallis, H.M., and A.H. Robinson (eds), *Cartographical Innovations: An International Handbook of Mapping Terms to 1900*, Map Collector Publications/International Cartographic Association, Tring, 1987.

Winchester, S., *The Map that Changed the World*, Penguin, London, 2002.

Woodward, D., (ed), *Art and Cartography: Six Historical Essays*, University of Chicago Press, Chicago, 1987.

PICTURE CREDITS

58 © British Antarctic Survey
59 © Oxford, Bodleian Library, C18 (14), plate 1
60 © Oxford, Bodleian Library, C18 (43), Sheet 7
61 Image based on Sheet 5.12, Fifth Edition of the IHO–IOC *General Bathymetric Chart of the Oceans* (*GEBCO*) Series, published by the Canadian Hydrographic Service, Ottawa, Canada, 1995
62 Swisstopo/ETH Archive, reproduced by permission of the Imhof family
63 Royal Collection Trust/© Her Majesty Queen Elizabeth II 2018
64 Reproduced by permission of Swisstopo, BA18071
65 Mick Ashworth
66 © Oxford, Bodleian Library, C17:70 Oxford (154)
67 Archaeological Museum, Istanbul, Photo © Zev Radovan/Bridgeman Images
68 © Oxford, Bodleian Library, MS. Douce 390, fols 5v–6r
69 Library of Congress, Geography and Map Division
70 © Oxford, Bodleian Library, E1 (160)
71 Wikimedia Commons
72 © Oxford, Bodleian Library, C18:45 Edinburgh a.1, sheet 8
73 © Oxford, Bodleian Library, Gough Maps 91, map 12
74 Harry Ransom Center, University of Texas at Austin, Kraus Map Collection, Kraus 13
75 © Oxford, Bodleian Library, Berkshire 6-inch, 1st edn, sheet 22
76 Courtesy of Giles Darkes
77 © 2016 London School of Economics and Political Science
78 © Oxford, Bodleian Library, I12 (281)
79 © Oxford, Bodleian Library, C18 b.3, plate 12
80 © Oxford, Bodleian Library, (R) B1 (199)
81 David Rumsey Map Collection, www.davidrumsey.com
82 Wikimedia Commons
83 Oxford University Museum of Natural History, WS/H/1/0/001
84 Courtesy of James A. Harrell
85 David Rumsey Map Collection, www.davidrumsey.com
86 David Rumsey Map Collection, www.davidrumsey.com
87 Rijksmuseum, Amsterdam, RP-P-1896-A-19368-2993
88 Utrecht University Library
89 Bibliothèque nationale de France
90 NOAA Office of Coast Survey/National Oceanic and Atmospheric Administration, U.S. Department of Commerce
91 © British Library Board. All Rights Reserved/Bridgeman Images
92 Library of Congress, Geography and Map Division
93 © Oxford, Bodleian Library, C1 (3) [1449] Sheet 57D N.E. parts 1&2
94 © Oxford, Bodleian Library, C21 (19B), sheet 31/22 SW
95 Wikimedia Commons
96 Boston Public Library, Norman B. Leventhal Map Center
97 Wikimedia Commons
98 © Oxford, Bodleian Library, Broxb. 67.8
99 © The Trustees of the British Museum, 1895,0112.1192-1197
100 © Oxford, Bodleian Library, C17:70 London (327)
101 U.S. National Parks
102 OS MasterMap, Courtesy of Ordnance Survey
103 Map data © 2018 GeoBasis-DE/BKG (© 2009 Google)
104 Windy.com
105 The Urban Forest and Ecology Team, Urban Sustainability, The City of Melbourne, www.melbourne.vic.gov.au
106 © Eric Fischer/Flickr
107 Reproduced courtesy of National Library of Scotland, base map © OpenStreetMap contributors
108 Used with permission. Copyright © 2017 Esri, ArcGIS Online, HERE, Increment Pm USGS, METi/NASA, and the GIS User Community. All rights reserved
109 © Oliver O'Brien/OpenStreetMap contributors

INDEX

References to images are in *blue*